Evolution Education and the Rise of the Creationist Movement in Brazil

Evolution Education and the Rise of the Creationist Movement in Brazil

Edited by Alandeom W. Oliveira
and Kristin L. Cook

LEXINGTON BOOKS
Lanham • Boulder • New York • London

Published by Lexington Books
An imprint of The Rowman & Littlefield Publishing Group, Inc.
4501 Forbes Boulevard, Suite 200, Lanham, Maryland 20706
www.rowman.com

6 Tinworth Street, London SE11 5AL, United Kingdom

British Library Cataloguing in Publication Information Available

Library of Congress Cataloging-in-Publication Data

Includes bibliographic references and index.

ISBN 978-1-7936-0148-3 (cloth)
ISBN 978-1-7936-0149-0 (Electronic)

Contents

Introduction

Historical Background and the Brazilian Educational Context

Alandeom W. Oliveira and Kristin L. Cook

We would like to begin this book by inviting the interested reader to carefully consider the following quotation:

> In the beginning there was only water, and above it the spirit of God. Then, God blew on the liquid surface and there appeared a light, which began to illuminate the darkness. This happened on the first day of creation. On the second day, He created the sky; on the third, fertile soil and plants, including fruit-bearing ones. On the fourth day, God created the Sun, the Moon, and the stars, placing them in their proper places on the cosmos. On the fifth day, fish in the water and birds in the sky were created. On the sixth day, God filled the earth with all species of animals and created man, to his image and resemblance. On the seventh day, he rested (p. 206).

It might come as a surprise to the reader, as it did to us, that the above quotation does not actually come from the Bible. It is in fact a passage from *Biologia em Contexto* (Amabis and Martho 2013), one of the most widely used biology textbooks in Brazilian public schools. The paragraph is on the first page of a chapter entitled "Chapter 9: Fundamentals of Biological Evolution," where one can also find a large and colorful reproduction of Michelangelo's *Creation of Adam* as portrayed in the Sistine Chapel's ceiling. This image occupies the entire upper half of the page, being followed by a fairly long section entitled "Myths of Creation" in which the authors provide a detailed account of creationism across ancient and medieval civilizations (Greece, Egypt, Inca, Viking, and Christian cultures). Students are introduced to creationism before even learning about evolutionary theory itself.

We came upon the above biology textbook while conducting a research about evolution education in Brazil (Oliveira and Cook 2018). As part of this study, we conducted a content analysis of the evolution chapters in five biology textbooks officially approved by the Brazilian government for use in public schools between the years of 2015 and 2017. Combined, nearly 1.7 million copies of these biology textbooks were distributed nationally free of charge, hence constituting some of the most widely used evolution curricula in Brazil. Our findings were both startling and puzzling. As science educators based in the United States, where the teaching of creationism in public school is deemed unconstitutional and a violation of the separation of church and state, it was rather surprising for us to find explicit and recurrent allusion to creationism in some of the most widely used evolution curricula in Brazil. It was particularly puzzling for us that the evolution units and chapters paid explicit attention to creationist ideas (both textually and visually) even though Brazilian public schools are said to be secular.

It was also noticeable to us that the above curricular trend has occurred amid a rapidly changing sociopolitical context characterized by considerable growth of anti-evolution attitudes among the populace and increasing political pressure to incorporate creationism in the public school classroom. A national survey recently revealed that, though evolution is accepted by more than half of the population (54 percent), the overwhelming majority of Brazilians (89 percent) now believe that creationism should be taught in schools, and that it should replace the theory of evolution in the school curriculum (75 percent) (Brum, Fonseca, and Cardoso 2005). Additionally, public defense of the teaching of creationism by high profile politicians (Silva and Prado 2010) and controversial legal rulings such as the Brazilian Supreme Court's recent approval of "confessional religious teaching"—focused on a single religion—have blurred the idea of secularism in public schools. Suggesting the rise of an influential creationist movement in Brazil, these numbers point to a unique and dynamic sociopolitical and educational context in a globalized era that allows for rapid worldwide travel of competing ideologies, unbounded by national borders.

The present book takes inspiration from these earlier findings. Their seemingly contradictory nature sparked our curiosity and motivated us to embark on a larger research effort aimed at better understanding how larger societal forces such as religion, law, media and politics have shaped the Brazilian educational landscape, and impacted the teaching and learning of evolution within an increasingly polarized educational discourse. We believe that a closer examination of how stakeholders in Brazil have coped with the pressure and navigated this controversy can help us better understand the dynamic sociological processes at play in the educational sphere of Latin American countries, as well as the interplay of societal forces behind current debates about the teaching and learning of evolution in public schools worldwide. As a first step in this intellectual journey, the next section provides historical background and contextual information about religion and public education in Brazil.

SECULAR EDUCATION IN A RELIGIOUS COUNTRY

Religiousness, particularly Catholicism, constitutes an important part of Brazilian subjectivity and culture. Studies in the field of psychology of religion emphasize the prominent role of religious experience and religious identity to Brazilians (Esperandio and Marques 2015). De Figueiredo-Cohen and Gvirtz (2009) take a step further by arguing that "In Latin America, the Catholic tradition is the basis of identity, originality and unity of the subcontinent" (p. 837). All of these scholars emphasize that religiosity and spirituality have a strong influence on cultural values, ways of thinking, and ways of being in Brazil (much like the rest of Latin America). Such a psychological trend persists despite Brazil's new religious configuration with dwindling numbers of Catholics and increasing presence of Evangelicals (particularly Protestants, Pentecostals, and Neo-Pentecostals). Census data shows that, between 1950 and 2010, Roman Catholicism in the country decreased from 93.5 percent to 64.3 percent of the population while Evangelicals increased to 22.2 percent (Follmann, 2017). Irrespective of religious affiliation, Brazilians generally continue to live and think spiritually (i.e., guided by a religious mental frame).

Far from new, Brazilians' religiousness can be traced back to colonial times. Historical accounts of the arrival of the Portuguese colonists in Brazil at the turn of the sixteenth century usually begin by emphasizing that their first action was to celebrate a Catholic mass in the newfound land. This important milestone in Brazilian history is portrayed by Victor Meirelles in his very famous painting *The First Mass in Brazil*, which is included in most school textbooks about Brazilian history. This painting underscores in a rather poetic and mythological way the centrality of religion to the nation's identity. In it, Brazil's birth is marked as a religious rite of passage similar to a baptism. This sort of historical imagery is commonly used to inculcate in Brazilian students from a very young age that their country began with a mass.

Throughout much of its history, Brazil was officially a Catholic State. The separation of church and state did not occur until the year 1891 with the constitutional adoption of the principle of state laicity (a translation of the French term laïcité that literally means "secularity"), which discourages religious involvement in government affairs and vice-versa. This in turn led to extended political debates about who should be in charge of education (De Figueiredo-Cohen and Gvirtz 2009). Pressure and influence from Catholic interest groups over the course of several decades eventually resulted in the inclusion of religious education as an optional discipline (an elective) in the new "secularized" school curriculum where it remains to this day. Its presence can be seen in current educational legislation in Brazil such as the *Law of Directives and Bases* (Brazilian House of Representatives 1996) which establishes that:

> Religious study, an optional discipline, is an integral part of the basic formation of the citizen and constitutes a discipline to be taught in elementary public schools, ensuring respect for the religious cultural diversity of Brazil, devoid of any form of proselytization (Art 33, § 1°).

This inclusion has been a source of controversy and political tension ever since, raising questions about who should teach it (state-certified instructors vs. church members), what content should be taught, and how to ensure respect for religious diversity and religious freedom. Such a legal provision was a direct result of the powerful Catholic Church's long history of political action, activism, and struggle to maintain some degree of control over education within the context of an increasingly laic Brazilian State. It was a problematic compromise that raised the possibility of there being multiple senses of State laicity, and opened the doors for longstanding political disputes over which parameters of secularization should be used to appropriately identify religiosity in public institutions. The complex state of affairs that resulted is described by Follmann and Gabatz (2017) as follows:

> Brazilian constitutional secularism, besides regulating the political action of secular religious groups at certain times, still does not have large legal or political devices to assist in the process of consolidating secularization and the consequent State laicity (p. 232).

Its ultimate effect was to blur political discourse with regard to secularism and laicity in ways that could be strategically exploited by religious interest groups to maintain a certain degree of influence in the Brazilian public sphere. The most recent chapter in this longstanding political struggle was the legal challenge placed against the implementation of confessional education in public schools.

CONFESSIONAL EDUCATION

In recent years, a legal battle ensued in Brazilian courtrooms over confessional religious education—a type of religious teaching limited in scope to a single doctrine. Rather than engaging in the more neutral study of world religions, students "learn about/from" one specific faith (Copley 2008). This bias toward a particular doctrine raises concerns about the possibility of students actually "learning into" a religion (being inducted or indoctrinated). Motivated by concerns over freedom of religion and arguing that such form of religious teaching is incompatible with a laic State, several education interest groups took the issue to court. Debates raged on in the media and in political spheres about what should or should not be happening in public schools, culminating in September of 2017 when the issue was taken up by the Brazilian Supreme Court. Deliberations were heated with several justices positing that the State should not simply neglect religious matters and that the elective nature of religious studies ensures freedom of religion since students have a choice to learn about a particular faith. A particularly memorable moment in the debate took place when Justice Gilmar Mendes (a supporter of confessional education) argued against what he considered to be just another "dictate of the politically correct":

> [If we continue to be politically correct] At some moment, will we start discussing removal of the Christ the Redeemer statue from Mount Corcovado, for symbolizing the Christian influence in our country? Or elimination of the national holiday celebrating Our Lady of Aparecida [patron saint of Brazil]? Or altering the names of our states? The State of São Paulo would become just Paulo? The State of Santa Catarina would become just Catarina? And how about the State of Espírito Santo [Holly Spirit], shall we consider calling it Espírito de Porco [colloquial expression in Portuguese meaning "Party Pooper"]?

Rhetorically speaking, Justice Mendes resorts to *reductio ad absurdum*, arguing that a completely secularized State would inevitably lead to absurd or ridiculous conclusions. For him the issue is just another instance of excessive political correctness, illogical in the sense that the Brazilian public sphere is historically and inherently Catholic, as evident in national monuments, national calendar, and Union states' nomenclature. His argument is consistent with curriculum studies showing that secular public schools in countries like the United States are not completely devoid of religion (Segall and Burke 2013). Though religion is not explicitly taught, various structural aspects of secular public education are inherently religious due to its Judeo-Christian roots, including the specialized language of schooling ("mission," "Cardinal Principles," "sabbatical"), educational symbols like the apple (used to represent teachers as well as Adam and Eve's sin), and the Christian calendar followed by schools (Burke and Segall 2011; 2015a; 2015b). To the astonishment of many, Justice Mendes's argument was successful and the final decision by the Supreme Court was in favor of the constitutionality of confessional religious education, thus ensuring its continuation in Brazilian public schools.

These historical events reveal a systematic effort to reconcile Catholic education with the increasingly secular Brazilian State. In an effort to rationalize the teaching of religion in public schools, Catholic educators redefined confessional education as a nonindoctrinatory public service central to the preparation of citizens for society (Carmody 2017; Copley 2008; Follmann 2017). To create compatibility with the secular social context, Catholic educators presented their confessional teaching simply as a public service enlightened by Christian principles, values, and practices professed by the Church (rather than an educational effort to proselytize the religion itself). Additionally, the legal mandate for complete secularization of the public sphere was questioned by many who were quick to point out that the Brazilian constitution itself makes reference to God as it establishes State laicity (the words "under the protection of God" can be found in its preamble). These Catholic groups' success in their political endeavor served as an important source of impetus for the eventual inclusion of creationism in biology textbooks as described at the onset of this chapter.

PHILOSOPHY EDUCATION

Another important educational development that had an impact on evolution education in Brazil was the recent integration of philosophy into the secondary public school curriculum. Brazilian educators and policy makers have long emphasized the importance of teaching and learning philosophy at the school level, highlighting its value as an essential component of schooling (Cesar 2012; Pinho 2014). This educational movement culminated in 2008 when philosophy became a mandatory subject in all grade levels of secondary education as both an integrated theme within content areas as well as a separate discipline with its own learning standards (PCN 2018). The *Law of Directives and Bases* (Brazilian House of Representatives 1996) identifies philosophical knowledge and abilities as an essential and required part of the school formation of new citizens. Even though the topic is actually absent from university entrance examinations, secondary students are expected to become familiar with the history of philosophy (from ancient to contemporary), understand important philosophical theories and schools of thought (Aristotle, Descartes, Locke, Kant, Hegel, Foucault, etc.), and practice how to philosophize (i.e., develop philosophical skills like reflection and questioning). Further, educators are called upon to create a transdisciplinary curriculum that seamlessly brings together philosophy and science. Secondary students should learn philosophy of science separately (in philosophy class) and while learning science.

Such a mandate had a strong impact on Brazilian science education, both in terms of educational policy and school curriculum. This is evident in the national science content learning standards for secondary schools, which require biology educators to integrate history and philosophy of biology into their teaching. When teaching the topic of evolution in particular teachers should promote student comprehension of the philosophical tenets of different explanatory systems as part of a historical approach. More specifically, students should learn that

> In the course of the history of humankind there have been several different explanations for the emergence and diversity of life, so scientific models have lived side by side with other explanatory systems, such as, for instance, those of a philosophical or religious inspiration (PCN 2018, p. 43).

These standards recommend that students need to learn how distinct historical periods and schools of thought held different ideas about the origin of life. In other words, evolution should be presented as one of many possible explanations for the origin of life (one of many philosophical thoughts or paradigms throughout human history).

Unsurprisingly, this historico-philosophical approach pervaded the government-approved biology textbooks examined as part of our previous study of evolution curriculum in Brazil (Oliveira and Cook 2018). In *BIO*, Lopes and Rosso (2013) included the following paragraph early in the evolution unit:

> Before life was understood as being the result of evolution, living beings were considered divine *creations*, referred as *creationism*. *Created* beings do not undergo changes over time, referred as fixism. Species were considered static groups of organisms, similar to an ideal and unchanging type, characterized by its own essence. Variations among members were merely accidental, and not essential. This way of thinking constitutes what it is known as essentialism. Essentialists do not accept variation, because for them variability is accidental and irrelevant (p. 10).

As can be seen above, Lopes and Rosso (2013) make explicit references to creationism, also identifying it as a conception of life that is aligned with a fixist philosophical paradigm no longer prevalent within scientific thought.

Likewise, in *Biologia Hoje,* Linhares and Gewandsznajder (2013) begin the chapter on evolution with the follow passage:

> According to prevalent thought until the XVIII century, each species had appeared in an independent fashion and retained the same characteristics. Even

> the Swedish naturalist Carl von Linné (1707–1778; Lineu in Portuguese), who created in 1735 the first system of biological classification, accepted this idea, known as *creationism* or fixism. In the beginning of the XIX century the hypothesis of transformation of species started to be defended by some scientists to explain diversity of species and the existence of fossils of organisms different from current organisms (p. 159).

Like the authors of other textbooks, Linhares and Gewandsznajder (2013) take a chronological approach first alluding explicitly to creationism, identifying it as part of fixism—a philosophical paradigm to which famous scientists like Linné once subscribed. Fixism is then contrasted to transformism, which is presented as a competing school of thought that emerged in a later century. "Where does life come from?" is treated as a philosophical question to which multiple answers have been formulated throughout the course of human history.

In Silva, Sasson, and Caldini's (2013) *Biologia*, the chapter on evolution begins with a large informative box entitled "Fixism and Transformism" that takes up the entire first page and top of the second page. It begins with the following passage:

> The notion that living beings change over time—called transformism—is for us as familiar as the idea that cells are the unit of life, or that that the DNA is our genetic material. However, for a long time, it was believed that each living species was fixed and unchanging—belief known as fixism—having appeared by *divine creation* and maintaining the same original characteristics. The fixist ideas, supported by Aristotle even in Antiquity, remained prevalent until the XIX century. This was due to the fact that Bible study reinforced the notion that the Earth had 6 thousand years, and that each life form was *constructed, one by one, by the Creator* (p. 202).

Once again, explicit allusion is made to creationism, which is presented as a fixist notion held at some point by famous scientists such as Linné and Georges. However, Silva, Sasson, and Caldini's (2013) account has a reversed timeline, focusing first on transformism and then going backward in time to older schools of thought. Biological change is again presented as one of multiple philosophical perspectives on the origin of life.

In sum, the legal mandate to integrate philosophy and science in Brazil provided an entryway for creationism to officially become part of the biology curriculum. Rather than being simply a topic covered in philosophy class or religious study, creationism is introduced as a biological concept in the science textbook, itself either explicitly or under the disguise of new labels such as "fixism" and "essentialism." Interestingly, relabeling was also an important part of the creationist movement in the United States where creationism was given the more neutral and exoteric name of "intelligent design." This similarity indicates that attribution of a new name/label is a tactic commonly used by creationists groups in different parts of the world to "secularize" religious notions such as divine creation as part of their educational reform efforts. Rhetorically speaking, the use of a new name with a less obvious religious embodiment allows creationists to diffuse the religious nature of their agenda and to a certain extent disguise creationism as a secular idea that can appropriately inhabit the science curriculum and classroom. This nominal tactic is consistent with Montgomery's (1996) contention that "the problem of naming cannot be divorced from the problem of knowledge" (p. 197) as nomenclature plays an important role in science; the act of (re)naming has the power of creating a new object for scientific study. Creationist groups' avoidance of more blatant religious terminology seems to constitute an effort to capitalize on this potential and strategically blur the line between religious and scientific knowledge.

CREATIONIST GROUPS

Another important contextual factor that should be taken into account is the increasing presence of creationist groups in Brazil. As debates raged on in the media and in political spheres about what should or should not be happening in public schools, creationist advocacy/political groups such as the *Brazilian Creationist Society*,[1] *Brazilian Institute of Intelligent Design*,[2] and the *Brazilian Association of Creation Research*[3] have been established. These creationist interest groups have actively campaigned for the teaching of creationism in publics, dramatically increasing their anti-evolution rhetoric in recent years. They tactically made available to the public a variety of spaces such as educational centers, websites, online stores, multimedia, annual conferences, et cetera, and ramped up their publication of anti-evolution propaganda like journals (e.g., the *Creationist* magazine) and books (e.g., *Biblical Creationist Cosmovision, Studies in Science and Religion*). Such tactics were very effective in increasing resistance to the teaching of evolution within the public educational system and producing a new generation of creationists in Brazil.

This apparent success has been highlighted by creationist groups based in the United States. For instance, Brazilian creationism has been the topic of several articles recently featured on *Evolution News and Science Today* (https://evolutionnews.org), an online periodical published by the Discovery Institute. Based in Seattle, this conservative and nonprofit institute has become notorious for its aggressive anti-evolution campaigns and efforts to promote the teaching of intelligent design and dissent from Darwinism. The

Discovery Institute was also a major force behind the "Dover Panda Trial" wherein the teaching of intelligent design in public school biology classes was challenged in federal court and eventually ruled to be unconstitutional and in violation of the legally mandated separation of church and state.

Interestingly, the Discovery Institute's newsflashes also reveal the existence of close ties between Brazilian and American creationist groups. In addition to portraying Brazilian creationism as "flourishing" and "shining" (Wells 2017), one article proudly announces the collaborative launching of a research center for creationist science at the Mackenzie Presbyterian University in São Paulo, a leading and traditional institution of higher education in Brazil (Klinghoffer 2017). These praises and joint venture suggest the possibility of influence and active involvement by this American institute in the creationist movement in Brazil. Having been unable to overcome legal hurdles in the United States, this institute appears to have to a certain extent refocused their efforts more broadly and globally, promoting and even joining educational efforts in other countries like Brazil.

DARWIN IN BRAZIL

Despite creationists' recent success in the educational arena, Darwin and evolution theory remain the subject of considerable admiration and celebration in Brazilian society. As in other parts of the world, Darwinian images are ubiquitous in Brazilian media including websites, cartoons, videos, and books. The same "cultish" admiration and veneration of Darwin reported by Browne (2009) also exists in Brazil. Admirers often refer to Darwin's visit to Brazil on his way to the Galapagos Islands in 1832, which is recounted in the detail by Darwin himself in his personal diary "The Voyage of the Beagle." Darwin's visibility in the Brazilian media was particularly prominent in 2009 when several events were held throughout the country to celebrate Darwin's 200th birthday and 150th anniversary of the publication of his book *On the Origin of Species*. These celebrations included a highly publicized visit by Sarah Darwin (Darwin's great-great-grand daughter) who, like him, traveled to the city of Salvador on a ship similar to the HMS Beagle for the filming of a Dutch documentary. Additionally, this bicentennial was marked by the publication of numerous books about Darwin's visit to Brazil when he was only twenty-two years old. One good example is the comic book "Darwin in Brazil" (de Almeida 2009) which provides a nonfictional account of Darwin's trip to Brazil on the Beagle in 1832. The book describes how Darwin attended the Mardi Grass festivities in the city of Salvador, lived in Rio de Janeiro, explored the tropical forest, and commented copiously on varied aspects of Brazilian culture and society like slavery. Also to celebrate Darwin's 200th birthday, a group of Brazilian researchers set out to precisely retrace and follow the footsteps of Darwin in and around Rio based on his notes (Amorim and Piolli 2009; Nogueira 2009). Funded by the Ministry of Science and Technology, this group of researchers sought to identify the exact roads traveled by Darwin and placed informative signposts in all the places he visited and stayed in (farms, houses, hotels, etc.).

EVOLUTION EDUCATION

Having examined several aspects of the country's sociopolitical and historical context, we now turn our attention specifically to the current state of evolution education in Brazil as revealed by recent research. By far, the most prominent theme in the existing literature is the need to improve biology teacher preparation programs in the country. While classroom-based studies are relatively rare, many researchers have examined biology teachers' conceptual knowledge and attitudes toward evolution instruction. These studies consistently indicate Brazilian biology teachers commonly hold misconceptions (creationist as well as Lamarckian) and misunderstandings of evolutionary theory (Caldeira, Araujo, and Carvalho 2012; El-Hani and Sepulveda 2010; Pazza, Penteado, and Kavalco 2010; Tidon and Lewontin 2004). Many are unsure how exactly evolution occurs (i.e., random effects, variation, and natural selection) whereas others conceive of evolution as being goal-driven.

In terms of attitudes, evidence also exists that many Brazilian biology teachers are amenable to teaching creationism as an alternative to evolution in their future classrooms (Penteado, Kavalco, and Pazza 2012), and some with a Protestant religious background are even inclined to repudiate scientific knowledge (El-Hani and Sepulveda 2010). These prevalent attitudes are not surprising given that some institutions like the Mackenzie Presbyterian University (a self-proclaimed confessional entity) go as far as incorporating creationism into the science classes taken by their pre-service teachers (Silva and Prado 2010).

Although classroom-based studies remain to be conducted, our examination of biology textbooks began to illuminate various facets of evolution instruction in Brazil. One noticeable curricular trend is that the early development of evolution theory is presented largely as a foreign endeavor carried out by European scientists (especially English and French). Most biology textbooks describe scientists like Darwin and Wallace as global travelers who at times visited Brazil for short periods of time to do some field work (collect specimens, make observations, etc.) and then returned to Europe where they eventually gained academic fame. In these highly Eurocentric and colonialist accounts, Brazil is simply a place from which evidence for evolution was extracted, or merely a quick stop along the journey without any direct relevance to the resulting scientific work. A good example of this type of narrative can be found in *Conexões com a Biologia*, where Bröckelmann (2014) writes:

> Charles Robert Darwin (1809–1882) was born in England . . . Darwin was the official naturalist of the expedition, in charge of observing and collecting specimens of plants and animals found in the places where the ship landed . . . during his visit to Brazil, which lasted four months, Darwin was in Bahia and Rio de Janeiro . . . Back *in England*, he posited that the individuals of a species compete for resources, especially food (pp. 98–99).

These historical narratives describe early work on evolution as being conducted exclusively by outside experts with no acknowledged form of participation or assistance from Brazilians. As a strictly European accomplishment, evolutionary theory's link to Brazil remains distant and mostly contextual. The seemingly colonialist agenda of these early European scientists prevents the emergence of any sense of Brazilian involvement or ownership.

In contrast, curricular descriptions of later scientific work tend to be more inclusive and open to Brazilian contribution. This is particularly true of the topic of human evolution. In most biology textbooks examined, the chapter on human evolution contains examples of research done in Brazil by Brazilian scientists, often in collaboration with foreign researchers. Nearly all textbooks include a picture of *Luzia, the Brazilian Cavewoman* (her cranium and reconstructed face). Discovered in the 1970s in the Brazilian archeological site of Lagoa Santa, Luzia was a young woman (*Homo sapiens*) between twenty and twenty-five years of age who lived in Brazil approximately 11,000 years ago. Her discovery is attributed to the Brazilian scientist Walter Neves at the University of São Paulo. One interesting exception can be found in Bröckelmann's (2014) *Conexões com a Biologia* which focuses instead on *Luzio, the Hunter* (named after Luzia due to their resemblance). In a section entitled "How a Pre-historic Paulista [native of Sao Paulo] lived," two whole pages are dedicated to the findings of a 2005 Brazilian study of a male Homo sapiens skeleton of approximately 10,000 years. In addition to describing these Brazilian discoveries, many textbooks also provide biographical blurbs of renowned Brazilian scientists (e.g., Niède Guidon) and descriptions of the country's major archeological sites (e.g., Serra da Capivara, Formação Santana). Though somewhat isolated and relatively infrequent, such inclusions help reshape the image of science presented to students as a national enterprise in which Brazilians are actively involved.

Another noticeable feature of Brazilian biology textbooks is the seemingly biased nature of the examples used to illustrate evolutionary concepts.

Consistent with previous research (Ballouard, Brischoux, and Bonnet 2011; Celis-Diez et al. 2016; Lindemann-Matthies 2005), biology curricula in Brazil tends to favor large, exotic, charismatic megafauna (e.g., Polar Bears, elephants, giraffes, lions, penguins, etc.) native to other countries over local and less popular animals such as insects in students' own backyards. The evolution chapters of the examined textbooks are for the most part dominated by images of mammals and birds. One good example is Mendonça's (2013) *Biologia* whose cover features a large and colorful picture of a Koala bear mother climbing a tree with a cub on her back. Nonetheless, Brazilian examples are not completely absent from this curricula. In Osorio's (2013) *Ser Protagonista Biologia*, the chapter "First Evolutionary Theories" begins with a large and bright photo of the Brazilian bird Urutau (*Nyctibius griseus*), who blends into tree branches and trunks much like a peppered moth. A few other examples involving Brazilian animals are also given in different parts of the text to illustrate various evolutionary concepts, including jaburu (Brazilian stork), capivara (large Brazilian rodent), and a less well-known Brazilian lizard (*Eurolophosaurus divaricatus*).

Lastly, problem sets at the end of evolution chapters usually include a few items that prompt students to apply evolutionary concepts to Brazilian scenarios or biological situations. One problem found in several books asks students to articulate a Darwinian explanation for the fact that Brazilian rattlesnakes (Crotalus spp.) have become quieter, increasing the risk of accidents for humans. A more detailed description of this problem can be found in Oliveira and Cook (2018).

The above curricular features appear designed to improve biology instruction by fostering local relevance, relatability, and contextualization. They help bring evolution closer to students' reality as a natural phenomenon that they can observe in their own backyards and of great importance to their daily lives (as opposed to an exoteric and inconsequential idea relevant to only distant foreign lands). These contextualizing strategies also help shape an image of science that is more inclusive and open to Brazilian participation (rather than just a foreign endeavor beyond the reality of Brazilians). Further research is needed to examine how and to what extent Brazilian biology teachers capitalize on these curricular resources.

BOOK FORMAT AND ORGANIZATION

It is against the historical and sociopolitical backdrop outlined above that the present book sets out to scrutinize the current state of evolution education and assess the apparent rise of the creationist movement in Brazil. To this end, we present a set of chapters that explore (1) evolutionary views, beliefs and understandings of various social groups in different parts of Brazil; (2) evolution curriculum and instruction across a variety of Brazilian schools and university settings; and, (3) how evolution education in Brazil compares to those of other countries and how its present state fares internationally. Together, these chapters illustrate a common, cross-cutting theme of the state of evolution teaching and learning in Brazil. Authors explore a range of perspectives on evolution from middle and high school students, university students, and the general population from different regions of Brazil as well as provide guidance for how to best support the teaching and learning of evolution in Brazil.

The book is divided into three main parts. Part I explores evolutionary views, beliefs and understandings in Brazil. These initial chapters empirically assess perceptions of high school students, university undergraduate students, and the general population of Brazilians in an effort to uncover how learners and citizens conceive of evolution and how it ought to be taught in schools. In "Rescuing Darwin in Brazil: How a General Population Sample Views the Teaching of Creationism and Biological Evolution," Heslley Machado Silva describes the views of more than four hundred Brazilian citizens on teaching evolution in schools and uncovers the pervasive view of citizens that creationism and evolutionism ought to be given the same time in the science classroom. Ideas for supporting science education in schools are presented. In "Creationists or Evolutionists? High School Students' Conceptions on the Origin and Evolution of Life," Pedro Teixeira examines ideas about the origin of life and change over time held by high school students from two public schools in the state of Rio de Janeiro. By investigating obstacles that may hinder students' development of sound scientific understandings, Teixeira reminds readers of the importance of honoring students' personal worldviews and belief systems while supporting their understanding of evolutionary theory. In all, these two chapters in part I begin to paint a fairly comprehensive picture of how Brazilians generally understand and feel about evolution and evolution education. Consideration is also given to possible ways that science educators can effectively deal with evolutionary understandings and feelings pervasive among the populace.

Part II explores evolution curriculum and instruction in Brazil. Diving deeper into pedagogical practices and teaching approaches related to evolution, the chapters in this section seek to elucidate what is happening in Brazilian classrooms through teacher surveys and empirical examinations of educational interventions. In "Brazilian High School Biology Teachers' Perception on Evolution and Its Teaching," Heslley Machado Silva and Eduardo Fleury Mortimer use the European questionnaire BIOHEAD-CITIZEN to survey sixty-two Brazilian high school biology teachers' difficulties and challenges when teaching about evolution. Teacher interviews reveal two commonly faced obstacles to the teaching of the subject, namely the beliefs of the teacher him/herself, and clash with the religiosity of the students. Recommendations are made on how to best support teachers. Similarly, Rita Tatiana Cardoso Erbs and Olma Karoline Cruz de Medeiros survey high school teachers' pedagogical strategies for teaching evolution and describe particular areas in which teachers identify the need for support. By analyzing autobiographical narratives of biology teachers who work in public schools in the city of Catalão, Goiás, the teachers' areas of emphases and their vulnerabilities in teaching evolution are explored. These two chapters add to the literature from American classrooms on teachers trying to grapple with the complexities of teaching evolution by defining the unique challenges and supports needed in Brazilian classrooms.

The next three chapters report intervention studies focused on specific pedagogical strategies used by Brazilian educators in various settings to teach evolution. In "Pedagogical Strategies for the Problem of Teleology in the Teaching of Biological Evolution," Beatriz Ceschim and Ana Maria de Andrade Caldeira examine the impact of a series of interventions on pre-service teachers' teleological conceptions. Based on their findings, the authors describe and evaluate pedagogical strategies that stimulated the learning of an evolutionary perspective more compatible with the scientific consensus. Further, the authors present a dialogic model that emphasizes student expression and promotes conceptual change by allowing students to directly challenge the misconception of teleological thinking. Gerda Maisa Jensen and Maria Elice de Brzezinski Prestes examine the design, implementation, and impact of a history of a science unit that uses primary literature (a chapter from Darwin's book) in the context of a literacy program for illiterate youth and adult learners in the city of São Paulo as part of a literacy program. The chapter describes various aspects of the intervention, including its contents, strategies and particular instructional materials developed following guidelines of inquiry teaching and problem-based activities. Data collected shows that the historical episode is a favorable path to promote the understanding of the concept of natural selection in youth and adult primary science education.

In "Challenges of Understanding Macroevolution among Brazilian Biology Students and Continuing Education Efforts," authors Leonardo Araújo, Ronaldo Paesi, and Voltaire Paes-Neto describe a university course designed to address misconceptions held by undergraduate biology students and support the development of scientifically sounder understandings of macroevolution in Southern Brazil. The authors' course includes visits to a paleontology museum, teaching tree thinking, and discussions regarding evolutionary developmental biology. Together, these chapters seek to provide readers a glimpse of existing practices and pedagogies commonly found in Brazilian classrooms. They also illuminate ways that Brazilian teachers navigate teaching about evolution as well as evidence of pedagogical strategies effectively used to target misconceptions and promote conceptual change and understanding of evolution.

Part III of the book presents international comparisons related to evolution with Brazil. Specific countries of comparison include Canada, Italy, Argentina, and the Galapagos Islands, all of which were intentionally selected to provide readers with unique contexts that share similar challenges and opportunities related to evolution education. In "What Is the Role of Sound Evidence in Evolution Education? A Research Program Following Students' Narratives in Brazil and Elsewhere," Nelio Bizzo investigates high school students' ideas about geologic timescale in the countries of Argentina, Italy, and Brazil. By doing so, the author seeks to better understand how cultural context shapes and impacts student ideas about evolution in particular ways.

Data from interviews and questionnaires of high school students bring to light explanations, which are in line with students' religious beliefs, with compelling narratives of factual knowledge seen under the lenses of local culture. Results reveal difficulties in understanding of the geological timescale. Extending from this study, in their chapter, "Comparative Studies of Students' Beliefs and Understandings in Brazil, Italy, and Galapagos Islands," Graciela S. Oliveira, Helenadja Santos Mota, Ana Maria Santos Gouw and Nelio Bizzo compare student conceptions in the context of places replete with fossil evidence for evolution. The authors also explore further possible correlations between different parameters, including religious beliefs in the three contexts. By doing so, the authors seek to determine the extent to which having a plethora of evidence within their local environments ("all around") affect these students' views of evolution/religion. It is reported that students who live in places with plenty of fossils do not necessarily show a higher understanding or belief in biological evolution.

Giuliano Reis, Adam Oliver Brown, Delano Silva, and Ana Julia Pedreira examine whether and how the topic of social Darwinism is addressed in commonly used science textbooks and taught by instructors in Brazil and Canada. In addition to providing a historical/philosophical account of social Darwinism, the authors seek to determine whether learners are provided with historical examples of evolution "misuse," and encouraged to critically consider the societal implications of evolutionary theory and its implications for helping make school science a site for resistance against prejudice and discrimination. The authors analyze classroom-based events and biology textbooks to explore the (mis)teaching of evolution as a phenomenon guided towards set standards of normalcy and naturalness. Combined, these international comparisons between Brazil and other countries help readers discern the ways in which context and culture play a significant role in student learning of evolution, and recognize the potential significance of issues raised and findings reported beyond Brazil itself.

Finally, in the last chapter, we provide a conclusion. More specifically, we articulate the main themes that emerged out of the preceding chapters, summarize how they can provide guidance on preparing and continuing to support science teachers and their students in the teaching and learning of evolution, and draw out implications (both theoretical and practical) for science teacher education. We then end by identifying areas in need of further research and practical consideration.

NOTES

1. http://www.scb.org.br/scb/
2. http://www.designinteligentebrasil.com.br/
3. http://abpc.impacto.org/

REFERENCES

Amabis, José Mariano, and Gilberto Rodrigues Martho. 2013. *Biologia em contexto: Adaptação e continuidade da vida* (1st ed., Vol. 2). São Paulo: Editora Moderna.

Amorim, Antônio Carlos, and Alessandro Piolli. 2009. "Bicentenário de Darwin: Comemorações confirmam vigor e atualidade do evolucionismo." *Ciência e Cultura* 61 no. 2: 62–63.

Ballouard, Jean-Marie, François Brischoux, and Xavier Bonnet. 2011. "Children prioritize virtual exotic biodiversity over local biodiversity." *PloS one* 6, n8: e23152.

Brazilian House of Representatives 1996. *Law of Directives and Bases* (# 9.394). Retrieved from http://www2.camara.leg.br/legin/fed/lei/1996/lei-9394-20-dezembro-1996-362578-normaatualizada-pl.html

Bröckelmann, Rita Helena. 2014. *Conexões com a biologia*. São Paulo: Editora Moderna.

Browne, Janet. 2009. "Looking at Darwin: Portraits and the making of an icon." *Isis*, 100, 542–570.

Brum, Eliane, Carlos Fonseca, and Nilson Cardoso. 2005. "E no princípio era o que mesmo? Época." Retrieved from http://revistaepoca.globo.com/Revista/Epoca/0,,EDG68197-6014,00-E+NO+PRINCIPIO+ERA+O+QUE+MESMO.html

Burke, Kevin, and Avner Segall. 2011. "Christianity and its legacy in education." *Journal of Curriculum Studies* 43, no. 5: 631–658.

Burke, Kevin, and Avner Segall. 2015a. "Teaching as Jesus making: The hidden curriculum of Christ in schooling." *Teachers College Record* 117, n. 3: 1–27.

Burke, Kevin, and Avner Segall. 2015a. "The religion of American public schooling: Standards, fidelity, and cardinal principles." *Journal of Curriculum and Pedagogy* 12, n. 1: 73–91.

Caldeira, Ana Maria de Andrade, Elaine Sandra Nicolini Nabuco de Araújo, and Graça Simões de Carvalho. 2012. "Creationism and evolution views of Brazilian teachers and teachers-to-be." *Journal of Life Sciences* 6: 99–109.

Carmody, Brendan. 2017. "The Catholic school: Non-confessional?" *International Studies in Catholic Education* 9, no. 2: 162–175.

Celis-Diez, Juan, Javiera Díaz-Forestier, Marcela Márquez-García, Silvia Lazzarino, Ricardo Rozzi, and J. J. Armesto. 2016. "Biodiversity knowledge loss in children's books and textbooks." *Frontiers in ecology and the environment* 14 n. 8: 408–410.

César, Renata Paiva. 2012. *O ensino de filosofia no Brasil.* Retrieved online from http://revistapandorabrasil.com/revista_pandora/renata/o_ensino_brasil.pdf

Copley, Terence. 2008. "Non-indoctrinatory religious education in secular cultures." *Religious Education* 103 n. 1: 22–31.

De Almeida, Flávio. 2009. *Darwin no Brasil.* Rio de Janeiro: Vieira & Lent.

De Figueiredo-Cohen, Maria, and Silvina Gvirtz. 2009. "The church and the state in Argentina and Brazil: Knowledge, religion, and pedagogy." In Robert Cowen, and Andreas Kazamias (eds.), *International Handbook of Comparative Education* (pp. 837–855). Netherlands: Springer.

El-Hani, Charbel, and Claudia Sepulveda. 2010. "The relationship between science and religion in the education of protestant biology preservice teachers in a Brazilian university." *Cultural Studies of Science Education 5* no. 1: 103–125.

Esperandio, Mary Rute Gomes, and Luciana Fernandes Marques. 2015. "The psychology of religion in Brazil." *The International Journal for the Psychology of Religion* 25: 255–271.

Follmann, José Ivo. 2017. "Brazil, Catholic religion and education: Challenges and prospects." *International Studies in Catholic Education* 9 no. 1: 76–88.

Follmann, José Ivo, and Celso Gabatz. 2017. "Secular state and freedom of religion in Brazil: The Brazil-holy see concordat and the 'general law of religions.'" *Ciências Sociais Unisinos* 53, no. 2, 225–233.

Klinghoffer, David. 2017. "Intelligent design shines in Brazil: More from discovery institute-Mackenzie launch." *Evolution News and Science Today.* Retrieved from https://evolutionnews.org/2017/05/intelligent-design-shines-in-brazil-more-from-discovery-mackenzie-launch/.

Lindemann-Matthies, Petra. 2005. "'Loveable' mammals and 'lifeless' plants: How children's interest in common local organisms can be enhanced through observation of nature." *International Journal of Science Education* 27, no. 6: 655–677.

Linhares, Sérgio, and Fernando Gewandsznadjer. 2013. *Biologia hoje: Genética, evolução, ecologia* (2nd ed., Vol. 3). São Paulo: Editora Ática.

Lopes, Sônia, and Sergio Rosso. 2013. *BIO* (2nd ed., Vol. 3). Brasilia: Editora Saraiva.

Mendonça, Vivian. 2013. *Biologia: O ser humano, genética, evolução* (2nd ed., Vol. 3). São Paulo: Editora AJS.

Montgomery, Scott. 1996. *The scientific voice.* New York: The Guilford Press.

Nogueira, Ítalo. 2009. *Grupo refaz passos de Darwin no Brasil*. Retrieved online from https://www1.folha.uol.com.br/fsp/ciencia/fe2303200801.htm

Oliveira, Alandeom, and Kristin Cook. 2018. "Evolution education and the rise of the creationist movement in Brazil." In H. Deniz and L. Borgerding (Eds.), *Evolution education around the globe* (pp. 119–136). Germany: Springer.

Osório, Tereza. 2013. *Ser Protagonista: Biologia* (vol. 3) (2nd ed.). São Paulo: SM.

Pazza, Rubens, Pierre Penteado, and Karine Kavalco. 2010. "Misconceptions about evolution in Brazilian freshmen students." *Evolution: Education and Outreach* 3: 107–113.

PCN 2018. *National curriculum parameters secondary education*. Brasília: Ministry of Education. Retrieved online: http://portal.mec.gov.br/seb/arquivos/pdf/pcning.pdf

Pinho, Romana Valente. 2014. "O ensino de filosofia no Brasil: Considerações históricas e político-legislativas." *Educação e Filosofia Uberlandia* 28, no. 56: 757–771.

Segall, Avner, and Kevin Burke. 2013. "Reading the Bible as a pedagogical text: Testing, testament, and some postmodern considerations about religion/the Bible in contemporary education." *Curriculum Inquiry*, 305–351.

Silva, César da, Sesar Sasson, and Nelson Caldini. 2013. *Biologia* (11th ed., Vol. 3). Brasilia: Editora Saraiva.

Silva, Heslley, and Isabelle Gonçalves de Oliveira Prado. 2010. "Creationism and intelligent design: Presence in the Brazilian educational policy." *Procedia Social and Behavioral Sciences* 2: 5260–5264.

Tidon, Rosana, and Richard Lewontin. 2004. "Teaching evolutionary biology." *Genetics and Molecular Biology* 27: 124–131.

Wells, Jonathan. 2017. Intelligent design is flourishing in Brazil. *Evolution News and Science Today*. Retrieved from https://evolutionnews.org/2017/12/intelligent-design-is-flourishing-in-brazil/.

Part I

EVOLUTIONARY VIEWS, BELIEFS, AND UNDERSTANDINGS IN BRAZIL

Chapter One

Rescuing Darwin in Brazil

How a General Population Sample Views the Teaching of Creationism and Biological Evolution

Heslley Machado Silva

Although Brazil is considered a secular country, a series of proposals regarding the teaching of Creationism has occurred amid changes in the religious spectrum. To find out how the population positions itself on the teaching of biological evolution and creationism, four questions from the Rescuing Darwin questionnaire was applied to 390 Brazilians. The results show that most samples are against the exclusive teaching of creationism in schools, unlike the majority of evangelicals and less educated participants who support it. However, most prefer creationism and evolutionism to be given the same time in the classroom, contradicting the idea of state secularism. This perception of creationism and evolution teaching was detected among respondents of all religious groups and across all levels of schooling. Nevertheless, most respondents believe that it would be prejudicial to science to remove the topic of evolution from schools, although fewer evangelicals and participants with less schooling agree. Finally, the majority thinks that it is possible to believe in both God and evolution, which is again less supported by evangelicals and the less educated. The findings highlight the importance of carefully considering the differentiated perceptions of evangelicals (the group that currently grows the most rapidly in Brazil compared to others) as well as the critical need to ensure that quality scientific education is offered to the general population.

The Darwinian theory of evolution was recently highlighted in the international media for its value and its impact on science as part of the celebrations of Charles Darwin's 200th birthday and the 150th anniversary of the publication of his fundamental book, *The Origin of Species* (Ayala and Avise 2009). At the same time, there is a counter-movement, a resurgence of creationism and its teaching in opposition to this scientific theory in various parts of the world (Scott 2009). More specifically, a sect of the creationist movement, known as Intelligent Design, seeks scientific recognition and media space, through an attempt to present itself as a bridge between science and religion (Shanks 2004). Despite lacking academic recognition, proponents of Intelligent Design have reached some relevance in various parts of the world, with intense action on the internet and even with the support of researchers with their seemingly conciliatory discourse (Silva 2017). From a Brazilian perspective, one can initially assume that this is a distant issue, resulting from the decades-long clash of opposing political groups in the United States. However, this conflict has arrived in Latin America, especially in Brazil (Cornish-Bowden and Cárdenas 2007). Faced with this new state of affairs and inspired by the British research Rescuing Darwin (Spencer and Alexander 2009), I sought to understand how a sample of the Brazilian population positions itself on the teaching of biological evolution and creationism.

BACKGROUND INFORMATION

The creationist movement has been active in Brazil in the political and educational spheres recently (Sepulveda 2004). Attempts of insertion with some success, at least in terms of media repercussion, have reached the Brazilian Congress and the legal environment (Martins 2001). In prestigious Brazilian schools and even universities, one can find examples of these trends in the substitution of pedagogical science materials about Darwinian biological evolution with religiously biased materials inspired and produced by North American creationists groups. Another example occurred in the state of Rio de Janeiro, in which a couple of former governors (known as the "Garotinhos") implemented a political and educational agenda that included the suggestion of creationism teaching in public schools and the hiring of teachers for the teaching of religion, a measure that went against the principle of secularism of the Brazilian State. These actions highlight the political muster and penetration of the most sophisticated version of creationism into Brazilian society, namely Intelligent Design (Martins 2013).

Although Brazil is predominantly a Catholic country, and this religion has in recent times been in favor of reconciliation with biological evolution theory and recognizing it as valid (Scott 2009), there have been complications. One of these complications is a strong conservative tendency present in certain Catholic groups currently in Brazil (Maués 2001). Another is the relevant change in the Brazilian religious spectrum. In recent years, there has been a decline in the percentage of Catholics and an increase in the number of evangelicals, many of whom are conservative and refractory to the theory of evolution. Moreover, there has been a slight increase in the number of atheists and agnostics (Silva and Mortimer 2014). This emerging Brazilian landscape is polarized. On the one hand, there are those who reject evolutionary theory, represented particularly by evangelicals and more moderately by Catholics. On the other side of this battlefield are agnostics and atheists, who aligned themselves with the scientific view and hold negative perceptions of evangelicals, who they see as being "backward" people. Such a state of affairs inevitably creates an environment of potential clashes in the classroom (Oliveira 2009).

METHODOLOGY

The research reported in this chapter was developed in Itaúna and Formiga, cities surrounding the metropolitan region of Belo Horizonte in the state of Minas Gerais (Midwest region). As part of this project, students of biological sciences at the University Center of Formiga (UNIFOR/MG) and at the University of Itaúna conducted interviews with the general population in the neighboring communities. The students received interview training and, subsequent to data collection, their interview data were combined and analyzed together. This allowed us to expand the area of coverage and achieve greater heterogeneity of participants.

The idea for this research originated from a large report, published in the *Folha de São Paulo*, the largest Brazilian newspaper, concerning the celebration of Charles Darwin's 200th birthday and 150th edition of his most important book, *The Origin of Species* (Colombo 2009). In this report, a British research, "Rescuing Darwin" (Spencer and Alexander 2009) is mentioned and five basic questions are presented, and several considerations are made about them and their results. In the report there are questions about what the results in Brazil of these questions would be. Then, the research group that I led at that time decided to answer these questions, on which the discussions in the report were made. In addition to these existing questions, seven more were included with the intention of focusing on the school issue on the teaching of creationism and evolutionism. We called this new questionnaire "Rescuing Darwin Brazil."

The proposal was to verify if the considerations made in the *Folha de São Paulo* newspaper report made sense in relation to the Brazilian population, or in a sample of this, and to discuss how this sample perceived issues related to the teaching of the same topics.

Although the questionnaire used in the "Rescuing Darwin Brazil" research consisted of twelve questions, many of which have already been analyzed in published articles. We have used only four of these questions for two reasons. First, they have already been discussed in previous analyzes (Leão et al. 2016a; Silva et al. 2015; Leão et al. 2016b). Second, because they deal with the teaching of creationism and evolutionism in Brazilian schools, and with the stance of the interviewees regarding these propositions.

Here are the twelve questions of the original Brazilian project. The four questions used in this chapter are highlighted in italics:

1. Intelligent design, the idea that evolution alone is not enough to explain the complex structures of some living things, so the intervention of a designer is needed at key stages.
2. The theistic evolutionism, the idea that evolution is the means that God used for the creation of all living things on earth.
3. The atheistic evolution, the idea that evolution makes belief in God unnecessary and absurd.
4. Young Earth Creationism, the idea that God created the world sometime in the last 10,000 years.
5. The thought of Darwin, the idea that humans evolved by a process of evolution which removes any need for God.
6. *On the idea that science classes should only address creationism, the idea that God created all living beings and they have not changed since then.*
7. On the proposal that science classes should only address how the origin of species and the evolution in the vision of Darwin's theory.
8. *About a suggestion that science classes should give the same value and use the same time to teach creationism and the theory of evolution.*
9. Science and religion can have a peaceful coexistence, even in an environment where ideas are so contrary as biological evolution and creationism.
10. *Some people argue that the teaching of evolution can be beneficial to the scientific development of humanity and deleting it would be detrimental.*
11. *It is impossible to believe in biological evolution and at the same time believe in God.*
12. What is your opinion about the teaching of evolution and creationism in schools? How would you prefer that your child be educated regarding this issue?

The questionnaire was given to 390 adults, with no criterion of choice. The sample can be described as random, or a probability sample, in that every participant has the same chance of being selected and interviewed, this random sample can be stratified into through variants, which make it possible to study subgroup or group characteristics in greater detail (Marshall 1996). The initial proposal that it should be a randomized sample has to be relativized in this case. Since biology students in two cities collected the data, it is probable that there was some deviation from the analyzed variables. For example, a higher education level than the Brazilian population in general because, when the questionnaire was submitted, the students likely sought out those to whom they would have easier access. Likewise, the geographic area of the research which involved, besides the cities where the universities were located, other surrounding municipalities. We sought to attenuate this deviation of the sample. In this chapter, the four questions that refer to suggestions for insertion of creationism teaching and the value of teaching of biological evolution are presented, analyzed, and discussed.

The response alternatives were presented in the same Likert scale model used in the British research. This more sophisticated type of prompt can measure level of agreement and disagreement of responses while excluding the simplicity of a yes-and-no answer (Mogey 1999). This scale will be defined from the following possible answers: (a) definitely true; (b) probably true; (c) probably untrue; (d) definitely untrue; (e) don't know. The variables education and religion of the respondents were considered for the analysis of the answers. A chi-square test (x^2) was used to verify if significant statistical differences occurred between the results found (Satorra and Bentler 2001). The sample was stratified by the variables schooling and religion. The chi-square test was used to verify if these variables were statistically significant in influencing the pattern of responses.

The numbers found in the sample (Table 1.1), demonstrate that the percentages are not demonstrative of the Brazilian population in general. Reinforcing

Table 1.1. Participants' demographic background.

		Sample	
		Number of respondents	*Percentage (%)*
SEX	Female	244	62.60
	Male	146	37.40
	Total Number of Respondents	390	100
RELIGION	Catholics	300	76.92
	Evangelicals	23	5.90
	Spiritualists	20	5.13
	Atheists	8	2.05
	No defined religion	28	7.18
	Other religions	11	2.82
	Total Number of Respondents	390	100
EDUCATION	Primary Education	6	1.54
	Elementary Education	39	10.00
	High School	105	26.92
	Incomplete Higher Education	104	26.67
	Higher Education	129	33.08
	Not answered	7	1.79
	Total Number of Respondents	390	100

the risk of a randomized sample ending up being influenced by the ease of access of the researchers. The respondents were mostly female. In relation to the religion, the majority was Catholic, as it is in Brazil largely. However, in relation to evangelicals, the sample presents a significantly lower percentage, and in relation to the Spiritists, the sample had a much higher index than the Brazilian population in general. Regarding schooling, the sample had a low percentage of individuals with low schooling and many with a higher level, different from the Brazilian population in general.

FINDINGS

In this section, we report our statistical results, focusing sequentially on each individual question from the questionnaire.

Question 6: Should Only Creationism Be Taught in Science Classes?

This question generated the following percentages of responses from participants: 11 percent definitely true; 15 percent probably true; 21 percent probably untrue; 53 percent definitely false; and, 1 percent don't know. These results suggest to a strong degree of rejection among participants, with 74 percent of the sample refuting the notion that only creationism should be taught in science classes and 26 percent supporting it. According to the chi-square test, it was observed that religion and school level influence the respondents' response ($p > 0.05$), the test demonstrated that these two factors are statistically significant for the percentages of responses related to this question.

A significant portion of this sample of Brazilians expressed an alignment with the hypothesis of Creationism in its most radical form, namely the Creationism of Young Earth (Silva, Mortimer, et al. 2015). However, when asked if it only creationism should be taught, there was disagreement. Another seeming contradiction was that, when asked about their acceptance of the Intelligent Design hypothesis, the idea of God guiding the evolutionary process, the level of acceptance among respondents is very high (Silva et al. 2016). Therefore, even though we have a degree of acceptance of the religious notions of Young Earth Creationism and Intelligent Design, more than what the original research was found in the United Kingdom (Spencer and Alexander 2009), the majority of this group of Brazilians was not enthusiastic about the suggestion of exclusively teaching creationism in science classes.

A possible explanation for this apparent contradiction is that Brazilians, by virtue of their religious tradition, tend to have a natural affinity for theories of the origin of life and man that involve the participation of a divine figure. However, these same Brazilians recognize the value of evolutionary theory, especially in the school environment. This positioning in relation to creationism in public schools can serve as a source of resistance to current attempts to insert the teaching of this religious aspect in the classes of science and biology that has occurred in Brazil, with projects in the Brazilian Congress and several proposals in the states, with emphasis on Rio de Janeiro state (Abrantes and Almeida 2006; Sepulveda 2004).

It is important to highlight the significance of this contradiction since Brazil is one of the countries in which creationism and Intelligent Design have been most successfully disseminated in recent years (Silva 2017). Intelligent Design posits that God has directed the evolutionary process and refutes many of the basic assumptions of Darwinian theory, such as random mutations, natural selection, and timing of evolution, promoting a poor mix of pseudoscience and religion (Shanks 2004). Therefore, the finding that there is no support for this more modern version of creationism to penetrate science classes can function as a healthy obstacle to any attempted insertion into the school curriculum. It is worth mentioning that this movement systematically lost legal battles to insert itself into public school education in the United States (Pennock 2003; Ayala 2008). It is hoped that, without the support of the population, this type of proposal will not prosper in Brazil and that it will not be necessary to reach out to the courts, as has happened with other forms of Creationism.

When considering the level of schooling of respondents, it was noticed that the larger the number of school years attended, the greater the respondents' rejection of the proposition of exclusively teaching creationism, especially at the level of high school and above. The level of agreement with this item decreased as respondents received more schooling.

Among interviewees with only primary and elementary education, the percentage of agreement with the suggestion of teaching creationism was high, around 80 percent and 70 percent of this sample, respectively. This result could be related to the lack of scientific knowledge on the subject of these individuals, because the contact with evolutionary theory, more forcefully or sometimes the only contact, occurs in High School (Porto and Falcão 2010). This conclusion must be viewed with reservations, as only one possibility, since to indicate the lack of schooling as a decisive factor for the acceptance or not of evolutionary theory, can demonstrate a degree of prejudice and determinism, that only those with a high level of schooling a can understand and not perceive other factors involved. Another aspect to be considered is that the theme of biological evolution is very scarce in elementary science textbooks in Brazil, making the topic distant from the science classes of many schools at this level (Tidon and Vieira 2009). It is a missed opportunity if this topic is not adequately addressed in elementary school science classes. Besides being one of the basic pillars of science and biology, students at this stage easily understand evolution when they study it in relation to fossils and common ancestry among species. Other studies demonstrate that this issue is not so simple, that the assimilation of something abstract like the evolutionary theory is not easy for the children, with a pattern of changes of concept with the increase of the age, starting from explanations of spontaneous generation, until arriving to definitely evolutionary or even creationist interpretations. The child's age, his knowledge of the natural history and the religion of his parents all seem to have a strong influence on this understanding and acceptance of evolutionary theory (Evans 2000). Another aspect that may be hindering the teaching of Darwinian evolution in Brazilian elementary education is a possible rejection of the topic by teachers and an alignment with creationism motivated by their religious beliefs and personal values (Araujo, Caldeira et al. 2009).

High school proved to be the most critical level of schooling for respondents' rejection of teaching creationism in lieu of evolutionism. From that schooling level on rejection was very high. This trend can be explained by two factors. The first one is that the Brazilian curriculum places more emphasis on the need to treat this topic specifically as part of biology teaching, as opposed to the more generalist curriculum of elementary school (Médio 2002). As a result of such disciplinary specificity, high school teachers tend to approach this topic more consistently. Another factor is that, for the general population, it is only at this educational level that the Darwinian theory is encountered academically. Most have contact with biological evolution only in the media in general, often in inaccurate and negative ways that are antagonistic to their personal religion and generate rejection (Silva et al. 2013).

Despite the increase in rejection of creationism in high school, almost 30 percent of those who completed this level agreed with the exclusive teaching of this religious perspective on the origin and evolution of species in class. One of the possible reasons, as reported by other studies, is the reduced time for teaching this topic in high school, which is often relegated to the end of the school year, if taught at all (Porto and Falcão 2010). Another factor to be considered is the nature of teachers' instructional approach to biological evolution, which may favor negation of the scientific vision of the evolutionary process. Two reasons for this inadequate approach can be suggested. Several studies support the hypothesis of limitations in teachers' knowledge, demonstrating that teacher training in Brazil presents problems, especially in regard to the topic of evolution. Moreover, these training problems are typically not remedied after teacher graduation and entry into the labor market due to a lack of continuous training that allows responding to these needs (Nuñez and Ramalho 2002; Oleques, Bartholomei-Santos et al. 2011). The second reason concerns internal religious and personal conflicts in relation to the teaching of biological evolution (Goedert, Delizoicov et al. 2003; Silva 2015).

Among those who made it to college or completed it, opposition to the exclusive teaching of creationism was quite high (more than 80 percent). These data show that a higher level of education coincides with a greater rejection of religious interference in formal education, which is expected in a secular country such as Brazil. However, it is not possible to say with absolute certainty that a longer period of education would prevent religious interference in teaching evolution as a Brazilian study has shown that students of different levels of schooling tend to accept evolutionary theory without the interference of their beliefs, as long as it has been adequately taught, at any level (Souza et al. 2009). Another study warns that despite accepting evolutionary theory, many students and teachers grasp this topic from a Lamarckian perspective (Carneiro 2004), that is, viewing evolution as driven by the law of use and disuse, by inheritance of acquired characteristics. Our survey data did not allow us to verify this possibility, because there was no question related to the Lamarckism.

When taking into account the religion variable, it is noticed that the evangelicals in our sample tended to agree with the proposition of the exclusive teaching of creationism (more than 60 percent). On the other hand, followers of Catholicism and of Spiritualism were those who had the greatest rejection of this suggestion, which is in agreement with the study that inspired this research (Colombo 2009; Spencer and Alexander 2009). It is important to consider this result in light of the recent changes in the Brazilian religious spectrum, which has seen a decrease in the number of Catholics, an increase in evangelicals, and growth of the political force of the latter group (Mariano 2013). Many evangelic sects are radical in their interpretation of the Christian Bible, increasing the chances of attempted interference to the teaching of biological evolution. This was evident in the several proposals reaching the Brazilian congress and the state of Rio de Janeiro (Machado and Mariz 2004).

A disturbing example of this evangelic movement is the position of the new minister of human rights in Bolsonaro's presidential government which began in 2019. This evangelical minister publicly declared to be contrary to the teaching of biological evolution, saying that evangelicals made a mistake by leaving science in the hands of scientists and allowing the teaching of Darwinian theory in schools (https://www.youtube.com/watch?v=ImgJ5c6-ock). It would be interesting to share with students and the general public in

Brazil that more progressive evangelical groups were able to reconcile their faith and the certainty that the evolutionary process happened, and that any interference in the educational sphere is counterproductive (Apple 2008).

Despite the growth of evangelicals, Brazil continues to be predominantly Catholic. In the sample investigated, only 23 percent of these agreed with the proposition concerning the teaching of creationism. Similarly, in a study with students of the third year of high school in Brazil, it was found that there was general acceptance of evolutionary theory among Catholics, without any major clashes between their religious beliefs and science. In the same study, it was shown that followers of this religion admit more easily to even believing in the origin of man through common ancestry (Madeira 2007). It is important to emphasize that this position of the majority of Catholics surveyed follows the official line of thought of the Catholic Church itself, which has adopted a stance of conciliation with the Darwinian theory. This stance dates back to several popes who shared a strong favorable attitude toward evolution, as can be seen in the declaration of Pope John Paul II:

> Today, almost half a century after the publication of the encyclical, new knowledge has led to the recognition of the theory of evolution as something more than just a hypothesis. It is indeed remarkable that this theory has been progressively accepted by researchers, following a series of discoveries in various fields of knowledge. The convergence, neither sought nor fabricated, of the results of work that was conducted independently is in itself a significant argument in favor of this theory. (Paul II 1996)

Participating Spiritists held a position similar to that of the Catholics. This religious group is the third with the largest number of followers in Brazil. The stance of followers of spiritism in relation to science and evolutionary theory is known for its flexibility. As predicted by the study that inspired this research, acceptance of evolutionism and the rejection of creationism was high for Spiritists (Colombo 2009). Curiously, when asked about the receptivity to Darwinism without consideration of faith, only taking into account the scientific view, the same sample of Spiritists expressed a high degree of rejection (Silva, Araújo et al. 2014). Finally, rejection of evolutionism by the Spiritists respondents was not paralleled by support of the teaching of creationism in Brazilian schools, which may hinder political and educational efforts in this sense.

Question 8: Should Religion and Science Be Given Equal Time in the Brazilian Classroom?

Concerning this question, the percentages of the responses were 46 percent definitely true; 27 percent probably true; 9 percent probably untrue; 14 percent definitely untrue; and 4 percent don't know. The chi-square test (x^2), showed that the variables religion and education were not statistically relevant to this question.

Unlike the previous survey item, this question suggests that creationism and evolutionism may share the same space within the classroom. When evolutionism was not excluded, the result was completely different, showing the strength of religiosity in this sample of the Brazilian population. This result does not correspond to what was suggested in the text that instigated this research, when comparing predicted results for Brazil with those of the United Kingdom. In that text, it was predicted that Brazilians would have less inclination to support religious proposals, but this question revealed a strong tendency to favor religious propositions in the sample investigated (Colombo 2009).

The view that science education should cover both creationism and evolutionism as a means to create supposedly plural and democratic education resonates not only with the religious but also with philosophers and humanists (Ruse 1982). However, researchers working in science education, educators, evolutionists, and scientists, in general, do not support this view (Scott 2009). The official Brazilian curriculum, proposed by the Ministry of Education, recommends the theme of the origin and evolution of life as the central theme of biology, which should be an articulator of other themes as the central and unifying element of this discipline (Brasil 2008) without religious interference. Nonetheless, many teachers of biology in Brazil still teach creationism in parallel to the theory of evolution or would like to teach it. Their argument for the teaching of the subject of evolution with religious commitment is an alleged effort to rend public education more pluralistic and allow open debate (Teixeira and Andrade 2014). No kind of creationism should be treated as a scientific hypothesis, nor should the version of intelligent design that calls itself a scientific theory be taught as if it were a viable theory parallel to biological evolution, as a possible alternative to it (Scott and Matzke 2007). Scientists concerned about evolution education have been forced to confront \intelligent design\ (ID).

None this proposal of equal time found support among the adherents of the sample of the three main Brazilian religions (Catholics, Evangelicals, and Spiritualists), with similar percentages. The theists of the sample had high percentages of agreement with the proposition that evolutionism and creationism should be given the same degree of relevance in science classes.

Curiously, among the three religions that have the greatest number of followers in Brazil, evangelicals showed the least level of agreement with this proposal. One possible explanation can be found in the prevalent response from members of this same religious group to the previous question, which proposed the exclusive teaching of creationism, receiving a high level of agreement. Therefore, it could be that radicalism regarding the teaching of creationism reduces evangelicals' level of acceptance of having a shared classroom space with the theory of evolution. Several researches in Brazil reinforce this possibility of radical positioning in relation to the evolutionary theory among several evangelical subgroups. Many believers in this religion consider it impossible to reconcile their faith with any type of evolutionism (Rodrigues and Clementino 2014; Mota 2013).

Members of the catholic religion were the ones who agreed the most with this question, showing some flexibility in relation to the evolutionist theory, compared to evangelicals. These findings are supported by other research of Catholic students, which has shown that their acceptance of Darwinian evolution is greater than that of evangelical students (Bizzo, Gouw et al. 2013). Despite being in decline in Brazilian society, Catholicism is still the most numerous religion, and it continues to have a strong influence in many spheres, including politics (Oro 2007). Therefore, in light of the results of questions 6 and 8, it can be assumed that proposals to include the teaching of creationism, but maintain the theory of evolution, may have greater acceptance in Brazil, especially among the majority religion. The British study that led to this research showed that most Catholics surveyed do not align themselves with the idea of atheistic evolutionism (Silva et al. 2014). Followers of this creed tend to stand on both science and religion like supporters of Intelligent Design and Theistic Evolution (Scott 2009).

Spiritists in the sample also had a favorable stance on the proposition of coexistence, being found at an intermediate level between evangelicals and Catholics. From these results it can be concluded that Brazilian theists support the teaching of creationism parallel to evolutionism, which should be viewed with concern, since this is one of the paths proposed by creationist groups who advocate intelligent design (Sepulveda 2004; Silva 2017).

The variable of education did not prove to be relevant in the consideration of the teaching of creationism alongside evolutionism, which was surprising. Regardless of the school level, whether or not they had had prior contact with the theory of evolution was no obstacle to accepting presence of a religious view alongside evolution in science classes. In question 6, discussed above, high school, a period of greater contact with Darwinian theory, saw a reduction in acceptance of the exclusive teaching of creationism. However, when respondents were presented with a more inclusive approach to science and religion, a similar pattern did not occur. These findings are consistent with response patterns elicited by the question about the acceptance of the intelligent design hypothesis. For many this hypothesis is seen as a compromise that effectively accommodates scientific and religious knowledge, even if it denies the basic assumptions of evolutionary theory by countless evidences and scientific proofs. The data from this question revealed that, regardless of the school level of participants in this Brazilian sample, there is a large degree of acceptance of intelligent design, and it increases among those with higher education (Silva et al. 2016).

It is a matter of concern that higher education did not interfere with acceptance of the idea that scientific and religious perspectives should be simultaneously treated in classrooms, demonstrating that even after having contact with evolutionary theory, these Brazilians did not consider it fundamental or thought it should be taught exclusively in science classes. It suggests that university instructors may have been unable to demonstrate the essential value of this theory for science as a whole, and especially for biology (Dobzhansky 1973). This mix of science and religion can be very harmful to Brazilian science education, since it could deprive the student of a more integrated approach to science and biology. Even among Brazilian undergraduates in biology, it was found that religious beliefs interfered in the understanding of Darwinian evolution, and may harm the formation of these future teachers. This in turn can have a ripple effect in science and biology classes and in the learning of their future students (Araujo et al. 2009). Instructors in Brazilian higher education institutions, have also been found to often teach the topic of evolution with conceptual misunderstandings, which creates problems in students' learning, fueling rejection of the theory and opening space for other explanations, mostly religious, for origin and diversity of species (Goedert, Delizoicov et al. 2003).

Question 10: Is Evolution Teaching Essential to Science?

Concerning question 10, which suggests that "Some people argue that the teaching of evolution can be beneficial for the scientific development of mankind and eliminating it would be harmful; you believe these arguments to be. . . .", the sample of Brazilians indicated the following percentages of responses: 66 percent definitely true; 23 percent probably true; 5 percent probably untrue; 4 percent definitely false; and, 2 percent don't know. Most of the sample expressed agreement, 89 percent agreed wholly or partially. This shows that the majority agreed with the academic field, which considers the topic of evolution fundamental to science (Dobzhansky 1973). These results seem to be in disagreement with what has been noted in other studies relative to issues of acceptance of the Young Earth creationism (Silva, Mortimer et al. 2015) and the intelligent design hypothesis (Silva et al. 2016), and rejection of evolutionism in its most materialistic form (Silva et al. 2014).

This apparent contradiction of aligning oneself with more than one form of creationism, rejecting the atheistic evolutionist, and considering the teaching of valiant biological evolution can be explained by the high religiosity of the sample, resulting in rejection of any approach that excludes the divine figure, while acknowledging that science dispenses with the knowledges of Darwinian theory. This view was confirmed in an extensive research, which detected the same perception among Brazilian students (Bizzo, Gouw et al. 2013).

These contradictions can be discussed from Barbour's (1990) work, which suggested four categories of relations between science and religion: conflict, independence, dialogue and integration. It may be argued that the category of independence gives a possible answer to the apparent contradiction of being a creationist who recognizes the value of teaching biological evolution. For most of our interviewees, it was possible to assimilate and live with a body of knowledge that to a certain extent conflicted with their faith. These participants recognized the importance of evolutionary theory while preserving their religious views.

Again, schooling was a positively relevant factor for the appreciation of Darwin's evolution theory, hence revealing how increasing numbers of school years allow for value recognition and demonstrating the critical role of science education in this regard (Wiles and Alters 2011). Those with lower levels of schooling had higher rejection levels or had difficulty expressing an opinion on the issue. As numbers of school years increased, the proportion of acceptance tended to also increase, although the statistical analysis was not significant ($p > 0.05$). Therefore, the influence of greater schooling on the acceptance or rejection of the premise was not clearly demonstrated by statistics, but a trend could be perceived, confirming what was detected in other questions about creationism and evolutionism with the same sample (Silva et al. 2014; Silva, Gibram et al. 2015). This trend corroborates the argument that, with a greater presence of biological evolution in the curriculum, there is a greater identification with the subject, especially in high school. Therefore it is imperative to increase this presence and improve the skills of biology teachers on the subject (Berkman and Plutzer 2012). However, it is interesting to reaffirm that the reported statistics did not reveal a significant difference, so it should be considered that even those with low level of education recognize, for the most part, the value of evolutionary theory.

Evangelicals showed relatively less agreement with this statement, but had a relatively high level of agreement overall, showing that even people in this more refractory group can come to recognize the value of the biological evolution through quality science education. This in turn demonstrates that there is no inherent incompatibility between science and religion in relation to the topic of evolution, as perceived by Darwin himself.[1] The study that led to this chapter considered that support for evolutionism would possibly be greater in Brazil than in the United Kingdom, partly due to the strong presence of Spiritualism and the tolerance of that religion in relation to science (Colombo 2009). In relation to this question this assumption was confirmed, because it was the Spiritists who most accepted the relevance of evolutionary theory. In contrast, Evangelicals were the ones who disagreed the most, demonstrating the resistance of this group, which has already been addressed. The Catholics presented an intermediate position, agreeing with what was found among students of this creed in Brazil (Bizzo, Gouw et al. 2013).

According to the 2010 Brazilian census, the religion-related groups that grow the most in the country are evangelicals and atheists. The former were more refractory to biological evolution, including education, whereas the latter are completely favorable to the exclusive teaching of Darwinian theory. Consequently, it can be foreseen that this debate on the teaching of evolutionary theory and its relevance will tend to intensify in Brazilian society and become polarized, with attempts to insert creationism and strong contrary reactions (Silva and Mortimer 2014). The sample confirms this trend, the chi-square test showed that the religion variable had a high impact on agreement or disagreement with the affirmation of the role of Darwinian theory in modern science suggested in the questionnaire ($p < 0.05$).

By combining data on religion and education, it can be noted that even among those in the sample who belong to religions more closely aligned with creationism, such as evangelicals, having higher levels of education makes it is possible to recognize the value of Darwinian theory for scientific development. It is then suggested that the category of independence between science and religion proposed by Barbour (1990) is evidenced, indicating that even deeply religious people, when adequately presented to evolutionary theory adequately in formal education or other can accommodate seemingly antagonistic visions, hence being able to recognize the value of each. This refutes one of the claims of the creationist movement and radical atheists, that Darwinian theory is incompatible with religion. These results show that, contrary to many creationists and proponents of intelligent design, biological evolution is not a theory in crisis, for in fact its value is increasingly recognized in the modern world (Coyne 2010).

Question 11: Is It Impossible to Believe in Both God and Evolution?

For the question 11, "Is it impossible to believe in biological evolution and at the same time believe in God?" About this affirmation, the majority of respondents from the sample (64 percent) tended to disagree; 52 percent disagree totally and 12 percent disagree partially. A total of 34 percent showed some degree of agreement, 21 percent agreed totally and 12 percent agreed partially. Only 2 percent did not know. Therefore, most of the sample disagrees that there is any obstacle that precludes accepting evolutionary theory and holding religious beliefs. These results reveal that for the majority in the sample, two or more understandings on a given topic can coexist (Caldeira, Araújo et al. 2011). This is consistent with conceptual change research showing that the individual does not necessarily have to abandon previous or alternative conceptions to assimilate a new scientific concept (Mortimer 1995).

Even though most of the sample identified themselves as being religious, this did not prevented evolutionary theory from being considered as true. Therefore, it is possible, even in environments with a strong religious influence, that people can construct an understanding of a controversial topic that allows assimilating the vision scientific (Cobern 1996). Thus, it is possible to deal with impasses when one crosses cultural boundaries, such as science and religion, since contradictory beliefs can coexist in the cognitive structure of the person (El-Hani and Bizzo 2002).

The four categories of Barbour (1990)—conflict, independence, dialogue, and integration—were used to analyze this issue of conceptualization. Most of the sample did not endorse the conflict thesis, preferring some dialogue or integration among different camps. There was also independence since it can be argued that science and religion could be seen as having different methods, themes, and languages that did not overlap or rival each other (Coutinho and Silva 2013).

In contrast, 34 percent of the respondents considered religious and scientific views to be incompatible. They agreed to some extent with the proposition that one must decide between two perspectives, between science and religion, since one conception could harm the other (Mahner and Bunge 1996). For this sample minority, the conflict thesis seems to prevail; for them, an impasse exists between scientific and religious knowledge (Barbour 1990).

Participants with more schooling considered the accommodation of religion and science within the subject of biological evolution to be possible, as also perceived by many believing scientists (Collins 2003). Responses to the "definitely true" option tended to decrease as the level of education increased from primary education to higher education. Furthermore, statistical analysis showed a significant shift between respondents with low levels of education (elementary and high school) and those with higher education (complete and incomplete higher education) ($p < 0.05$). These data indicate that schooling influences the compatibility between religion and the theory of evolution. However, statistical results for question 10 showed that even those who have had a limited formal education understood the importance of evolutionary theory.

High school appears to be "the great divide" between agreement and disagreement with this question. High school contact with evolutionary theory allows one to perceive that it is not incompatible with faith. There seems to be a change in the conceptual structure that allows the individual to notice that two types of concepts can exist, even in apparently antagonistic questions, without a conceptual change being necessary (Mortimer 1995). This is consistent with studies showing that, in the educational environment, there can be a change from a fundamentalist to a more liberal one, allowing the accommodation of scientific and religious knowledge (Bizzo, Gouw et al. 2013; Sepulveda and El-Hani 2004).

In Brazil, pre- and in-service biology teachers recognize evolution and natural selection as part of the process of species formation, but most believe that there is a "breeder" who governs this process (Caldeira, Araújo et al. 2011). Another Brazilian study of an upper biological course described a process of gradual accommodation between science and religion regarding evolutionary theory, and an adaptation to the new knowledge without any loss of religiosity (Silva, Araújo, et al. 2014).

Despite some decrease in agreement as education increased in the sample of Brazilians, a significant portion of respondents maintained the view of conflict between science and religion (Barbour 1990). This highlights the need to acknowledge that, if a new conception causes conflict over a previous one, there may be a refusal of conceptual change or a substitution to accommodate the new concept (Posner et al. 1982). This maintenance of the conflict may also be related to the literal interpretation of the religious doctrines and of the Holy Scriptures, which would make any conceptual change related to this question difficult (Mahner and Bunge 1996). Another possibility for this resistance to change or accommodation could be related to the limited time given by many biology teachers to the topic of evolution, and the assumption made by many that these teachers do not believe in Darwinian theory, which ultimately influences their teaching and, consequently, student learning (Berkman and Plutzer 2012). Finally, it is necessary to question the teachers' background on the topic of evolution and its teaching, which reverberates in its acceptance and in the way of teaching it, potentially impacting the classroom treatment and having repercussions for the assimilation of the student (Rutledge and Mitchell 2002).

When one considers the variable of religion, mostly evangelicals and atheists found it impossible to reconcile evolution with religion for different reasons, hence providing a prelude to the types of conflicts that may occur in the classroom between these two groups (Silva 2015), which will require, of the Brazilian teachers, new abilities to deal with conflict and to have conciliatory attitudes through alternative approaches. Among the three main Brazilian religions, the Spiritists were those who were the furthest away from the category of conflict (Barbour 1990). This finding corroborated what was predicted by the British study that inspired this chapter, namely that spiritism would be more tolerant and less conflicting with science (Colombo 2009). Catholics held an intermediate position. It seemed that the most intense religiosity, most common among evangelicals, was antagonistic only when one assumed that evolutionism had an atheistic presupposition (Silva et al. 2013). Empirical evidence from the scientific field has be shown to have little or no impact on the fervently religious (Sanz 2013). The result is a barrier to the assimilation and accommodation of evolutionary theory. However, in the case of the Brazilian evangelicals in the sample, even though they were religiously enthusiastic in its majority, a significant portion could reconcile the two fields of knowledge in relation to the Darwinian theory.

Again, two antagonistic groups in relation to belief had similar results, obviously for different reasons. Atheists and evangelicals had more than 50 percent agreement with the information (Oro 2007; Mariano 2013). With the growth of these two groups, there is a risk that the category of conflict (Barbour 1990) may prevail over the possibility of believing in both God and assimilating evolutionary theory. To many followers of the evangelical religion, the Bible represents the word of God, the message is direct, word for word. Therefore, any information or discovery that contradicts God's words must be rejected and fought. With no room for interpretation, clash is inevitable (Bizzo 2013). To prevent this conflict from increasingly occurring, it is essential that classroom-based approaches of mutual respect be pursued in biology classes. However, excessive emphasis on respect ignores the need to treat biological evolution without renouncing any of the discoveries and evidences of this field. When teaching evolution, it is important to clearly demonstrate that even deeply religious individuals can be scientists, users of scientific methodology, and knowledgeable and users of Darwinian theory in their research (Collins 2003).

The findings show that reduced education and evangelical religion were important factors in the rejection of evolutionary theory and acceptance of creationism. Nonetheless, with quality scientific education that demonstrates the academic and human value of the topic of biological evolution, it might be possible to engender change in the Brazilian religious profile.

CONCLUSION

Responses to sample questions have shown that teaching biological evolution and inserting creationism in schools are complex topics, and that when individuals conceive of these topics, perceptions are often seemingly contradictory. In general, Brazilian respondents did not align themselves with the idea of exclusive teaching of creationism but appreciated the teaching of both conceptions at school, recognized the value of Darwinian theory, and consider it possible to accept it and believe in God.

From this analysis, one can conclude that neither religion nor limited schooling constitute an unsurmountable obstacle to the assimilation of evolutionary theory, although they are difficult. Based upon these results, it can be proposed that teachers need to avoid a clash between science and religion when teaching the topic of biological evolution in order to avoid rejection and facilitate assimilation. Additionally, biology teachers need access to continued professional development on the subject. Being abreast of new discoveries in the field will enable biology teachers to more effectively demonstrate the value and applicability of evolution to students' lives.

NOTE

1. *It seems to me absurd to doubt that a man may be an ardent Theist & an evolutionist.* Darwin Correspondence Project, "Letter no. 12041," accessed on 22 September 2018, http://www.darwinproject.ac.uk/DCP-LETT-12041.

REFERENCES

Abrantes, Paulo, and Fábio Portela Lopes de Almeida. 2006. "Criacionismo e Darwinismo Confrontam-Se Nos Tribunais . . . Da Razão e Do Direito." *Episteme* 11 (24): 357–402.

Apple, Michael W. 2008. "Evolution versus Creationism in Education." *Educational Policy* 22 (2). Sage Publications: Los Angeles, CA: 327–35.

Araujo, Elaine Sandra Nicolini Nabuco, Ana Maria de Andrade Caldeira, João José; Caluzi, and Graça Simões Carvalho. 2009. "Concepções Criacionistas e Evolucionistas de Professores Em Formação e Em Exercício." In *VII Enpec–Encontro Nacional de Pesquisa Em Educação Em Ciências*, 12. Florianópolis.

Araújo Porto, Paulo Roberto de, and Eliane Brigida Morais Falcão. 2010. "Teorias Da Origem e Evolução Da Vida: Dilemas e Desafios No Ensino Médio." *Ensaio Pesquisa Em Educação Em Ciências* 12 (3): 13.

Ayala, Francisco J. 2008. "Science, Evolution, and Creationism." *Proceedings of the National Academy of Sciences of the United States of America* 105 (1): 3–4. https://doi.org/10.1073/pnas.0711608105.

Ayala, Francisco J, and John C Avise. 2009. "Darwin at 200." In *Proceedings of the National Academy of Science*, 106:2475–76.

Barbour, Ian G. 1990. *Religion in an Age of Science*. Harper & Row: San Francisco.

Berkman, Michael, and Eric Plutzer. 2012. "An Evolving Controversy: The Struggle to Teach Science in Science Classes." *American Educator* 36 (2). ERIC: 12.

Bizzo, N, Ana Maria Santos Gouw, and Helenadja Mota Rios Pereira. 2013. "Evolução e Religião: O Que Pensam Os Jovens Estudantes Brasileiros." *Ciência Hoje* 50 (300): 26–31.

Bizzo, Nelio Marco Vincenzo. 2013. "Criacionismo versus Evolucionismo: Literalismo Religioso e Materialismo Darwiniano Em Questao." *Filosofia e História Da Biologia* 8 (2). Associação Brasileira de Filosofia e História da Biologia-ABFHiB: 301–39.

Brasil, Ministério da Educação. Secretaria da. 2008. *Orientações Curriculares Para o Ensino Médio*. Ministério da Educação.

Caldeira, Ana Maria de Andrade, Elaine Sandra Nicolini Nabuco de Araújo, and Graça Simões de Carvalho. 2011. "Brazilian Teachers' Conceptions about Creationism and Evolution." In *Authenticity in Biology Education: Benefits and Challenges: A Selection of Papers Presented at the 8th Conference of European Researchers in Didactics of Biology (ERIDOB)*, 309–21. Universidade do Minho. Centro de Investigação em Estudos da Criança (CIEC).

Carneiro, Ana Paula Netto. 2004. "A Evolução Biológica Aos Olhos de Professores Não-Licenciados." Universidade Federal de Santa Catarina.

Cobern, William W. 1996. "Worldview Theory and Conceptual Change in Science Education." *Science Education* 80 (5). Wiley Online Library: 579–610.

Collins, Francis S. 2003. "Faith and the Human Genome." *Perspectives on Science and Christian Faith* 55 (3): 142–53. http://www.asa3.org/ASA/PSCF/2003/PSCF9-03Collins.pdf.

Colombo, Sylvia. 2009. "Darwin Nas Mãos de Deus." *Folha de São Paulo*, 4.

Cornish-Bowden, Athel, and Maria Luz Cárdenas. 2007. "The Threat from Creationism to the Rational Teaching of Biology." *Biological Research* 40 (2). Sociedad de Biología de Chile: 113–22.

Coutinho, Francisco Ângelo and Fábio Augusto Rodrigues Silva. 2013. "Ciência e Religião: Uma Guerra Desnecessária." *Ciência Hoje* 51 (304): 18–21.

Coyne, Jerry A. 2010. *Why Evolution Is True*. Oxford University Press.

Dobzhansky, Theodosius. 1973. "'Nothing in Biology Makes Sense except in the Light of Evolution.'" *The American Biology Teacher* 35 (125–29). https://doi.org/10.2307/4444260.

El-Hani, Charbel Niño, and Nélio Marco Vicenzo Bizzo. 2002. "Formas de Construtivismo: Mudança Conceitual e Construtivismo Contextual." *Ensaio Pesquisa Em Educação Em Ciências* 4 (1): 37–58.

Evans, E Margaret. 2000. "The Emergence of Beliefs about the Origins of Species in School-Age Children." *Merrill-Palmer Quarterly (1982-)*. JSTOR, 221–54.

Goedert, Lidiane, Nadir Castilho Delizoicov, and V. L. Rosa. 2003. "A Formação de Professores de Biologia e a Prática Docente-o Ensino de Evolução." *Atas Do IV Encontro Nacional de Pesquisa Em Educação Em Ciências (ENPEC). Bauru-SP: ABRAPEC.*

Leão, Isabela Maria Silva, Heslley Machado Silva, and E. S. N. N. de Araujo. 2016a. "Must Evolutionism and Creationism Have the Same Space in Science Classes." *Journal of Teaching and Education* 6 (1): 171–78.

Leão, Isabela Maria Silva, Heslley Machado Silva, Elaine Sandra Nicolini Nabuco, de Araujo. 2016b. "Teaching of Creationism: Analyzing the Opinions of a Sample of Brazilians." *Journal of Teaching and Education*, no. 01: 2165–6266. http://universitypublications.net/jte/0501/pdf/DE5C294.pdf.

Machado, Maria das Dores Campos, and Cecília Loreto Mariz. 2004. "Conflitos Religiosos Na Arena Política: O Caso Do Rio de Janeiro." *Ciencias Sociales y Religión/Ciências Sociais e Religião* 6 (6): 31–49.

Madeira, Andréia Porto Luiz. 2007. "Fé e Evolução: A Influência de Crenças Religiosas Sobre a Criação Do Homem Na Aprendizagem Da Teoria Da Evolução Com Alunos Do 3° Ano Do Ensino Médio. 2007. 186 F." Dissertação (Mestrado–Programa de Estudos Pósgraduados em Ciência da. . . .

Mahner, Martin, and Mario Bunge. 1996. "Is Religious Education Compatible with Science Education?" *Science & Education* 5 (2). Springer: 101–23.

Mariano, Ricardo. 2013. "Mudanças No Campo Religioso Brasileiro No Censo 2010." *Debates Do NER* 2 (24): 119–37.

Marshall, Martin N. 1996. "Sampling for Qualitative Research." *Family Practice* 13 (6). Oxford University Press: 522–26.

Martins, Maurício Vieira. 2001. "On Darwin, Black Boxes and the Amazing Return of 'creationism.'" *História, Ciências, Saúde-Manguinhos* 8 (3). SciELO Brasil: 739–56.

———. 2013. "Quando Uma Sociologia Da Ciência Se Faz Necessária: Aspectos ContemporâNeos Do Embate Entre Criacionistas e Evolucionistas." *Filosofia e História Da Biologia* 8 (2). Associação Brasileira de Filosofia e História da Biologia-ABFHiB: 279–99.

Maués, R Heraldo. 2001. "Tradição e Modernidade Conservadoras No Catolicismo Brasileiro: O Apostolado Da Oração e a Renovação Carismática Católica." *Sociedad y Religión* 22: 23.

Médio, P. C. N. Ensino. 2002. "Orientações Educacionais Complementares Aos ParâMetros Curriculares Nacionais–Ciências Da Natureza, Matemática e Suas Tecnologias." *Brasília: MEC/Semtec.*

Mogey, Nora. 1999. "So You Want to Use a Likert Scale." *Learning Technology Dissemination Initiative* 25.

Mortimer, Eduardo F. 1995. "Conceptual Change or Conceptual Profile Change?" *Science & Education* 4 (3). Springer: 267–85.

Mota, H. S. 2013. "Evolução Biológica e Religião: Atitudes de Jovens Estudantes Brasileiros 'Biological Evolution and Religion: Attitudes of Young Brazilian Students.'" Dissertação (Doutorado em Educação)—Universidade de São Paulo, São Paulo.

Nuñez, Isauro Beltrán, and Betânia Leite Ramalho. 2002. "Estudo Da Determinação Das Necessidades de Professores: O Caso Do Novo Ensino Médio No Brasil–elemento Norteador Do Processo Formativo (Inicial/Continuado)." *Revista Iberoamericana de Educación* 29 (1): 1–18.

Oleques, Luciane Carvalho, Marlise Ladvocat Bartholomei-Santos, and Noemi Boer. 2011. "Evolução Biológica: Percepções de Professores de Biologia." *Revista Electrónica de Enseñanza de Las Ciencias* 10 (2): 243–63.

Oliveira, Graciela da Silva. 2009. "Aceitação/Rejeição Da Evolução Biológica: Atitudes de Alunos Da Educação Básica." Universidade de São Paulo.

Oro, Ari Pedro. 2007. "Religião, Coesão Social e Sistema Político Na América Latina." São Paulo, IFHC.

Paul II, Pope John. 1996. "Truth Cannot Contradict Truth." *L'Osservatore Romano* 30.

Pennock, Robert T. 2003. "Creationism and Intelligent Design." *Annual Review of Genomics and Human Genetics* 4 (1): 143–63.

Posner, George J, Kenneth A Strike, Peter W Hewson, and William A Gertzog. 1982. "Accommodation of a Scientific Conception: Toward a Theory of Conceptual Change." *Science Education* 66 (2). Wiley Online Library: 211–27.

Rodrigues, Wellington Gil, and Priscila Prazeres Clementino. 2014. "O Ensino Da Teoria Evolucionista Na Perspectiva Dos Professores de Ciências Da Rede Adventista de Ensino." *Revista Formadores* 7 (3): 5.

Ruse, Michael. 1982. *Darwinism Defended.* Addison-Wesley, Advanced Book Program/World Science Division.

Rutledge, Michael L, and Melissa A Mitchell. 2002. "High School Biology Teachers' Knowledge Structure, Acceptance & Teaching of Evolution." *The American Biology Teacher* 64 (1). BioOne: 21–28.

Sanz, Vicente Claramonte. 2013. "La Evolución Del Creacionismo: Del Mito Cosmogónico a La Pseudociencia Biológica." *Filosofia e História Da Biologia* 8 (2). Associação Brasileira de Filosofia e História da Biologia-ABFHiB: 361–79.

Satorra, Albert, and Peter M Bentler. 2001. "A Scaled Difference Chi-Square Test Statistic for Moment Structure Analysis." *Psychometrika* 66 (4). Springer: 507–14.

Scott, Eugenie C. 2009. *Evolution vs. Creationism: An Introduction.* Vol. 62. Univ of California Press.

Scott, Eugenie C, and Nicholas J Matzke. 2007. "Biological Design in Science Classrooms." *Proceedings of the National Academy of Sciences of the United States of America* 104 Suppl: 8669–76. https://doi.org/10.1073/pnas.0701505104.

Sepulveda, Claudia. 2004. "Movimento Criacionista: Um Risco à Formação Científica e Cultural Dos Alunos Da Rede Pública Carioca." *Boletim Da SBEnBio* 1: 4–9.

Sepulveda, Claudia, and Charbel Niño El-Hani. 2004. "Quando Visões de Mundo Se Encontram: Religião e Ciência Na Trajetória de Formação de Alunos Protestantes de Uma Licenciatura Em Ciências Biológicas." *Investigações Em Ensino de Ciências* 9 (2): 137–75.

Shanks, Niall. 2004. *God, the Devil, and Darwin: A Critique of Intelligent Design Theory.* Oxford University Press.

Silva, Heslley Machado, Elaine Nicolini Nabuco deAraújo, and Paloma Rodrigues Silva. 2014. "Perception of Brazilian Citizens about Darwinism and the (Non) Influence of God in the Evolutionary Process." Istanbul: Proceedings of INTCESS14—International Conference on Education and Social Sciences P. https://www.researchgate.net/publication/275100477.

Silva, Heslley Machado. 2015. "Professores de Biologia e Ensino de Evolução: Uma Perspectiva Comparativa Em Países Com Contraste de Relação Entre Estado e Igreja Na América Latina." Tese apresentada a Faculdade de Educação da Universidade Federal de Minas Gerais. Belo Horizonte.

———. 2017. "Intelligent Design Endangers Education." *Science* 357 (6354): 880.1-880. https://doi.org/10.1126/science.aao3245.

Silva, Heslley Machado, Graça S Carvalho, Paloma Rodrigues da Silva, Daiana, Evilin Gibram, Estadual Paulista, and São Paulo. 2016. "A Percepção Sobre a Hipótese Do Design Inteligente No Brasil A Percepção Sobre a Hipótese Do Design Inteligente No Brasil (Minas Gerais)." *Revista Conexão Ciência I* 11.

Silva, Heslley Machado, Daiana Evilin Gibram, Rosa Branca Tracana, and Graca S. Carvalho. 2015. "Can One Accept the Theory of Evolution and Believe in God as Well?" *Procedia–Social and Behavioral Sciences*. https://doi.org/10.1016/j.sbspro.2015.07.178.

Silva, Heslley Machado, and E. F. Mortimer. 2014. "Rescuing Darwin Brazil." In *Latin American Perspectives on Science and Religion. International Perspectives on Science, Culture and Society*, edited by Ignacio Silva, 1st ed., 1:208. London: Routledge.

Silva, Heslley Machado, Fleury Mortimer, Elaine Sandra Nicolini, Ana Carolina, and Lopes Souza. 2015. "A Percepção no Brasil sobre a Hipótese Do Criacionismo Da Terra Jovem." *Quaerentibus* 3 (5): 7–26.

Silva, Heslley Machado, Elaine Sandra Nicolini Nabuco de Araújo, Daiana Evilin Gibram, and Graça S. Carvalho. 2014. "Conceptual Change about Evolution and Origins of Life throughout an Undergraduate Course of Biological Sciences."

Silva, Heslley Machado, Paloma Rodrigues da Silva, Ana Carolina Lopes Souza, and Elaine Sandra Nicolini Nabuco de Araújo. 2013. "A Influência Da Religiosidade Na Aceitação Do Evolucionismo: Um Estudo Em Uma Amostra Da População Brasileira." *Conexão Ciência* (Online). https://doi.org/10.24862/cco.v8i1.164.

Souza, R. F. de M. de Carvalho, Tiemi Matsuo, and Dimas A M Zaia. 2009. "Evolucionismo x Criacionismo: Aceitação e Rejeição No Século 21." *Ciência Hoje* 43 (256): 36–45.

Spencer, Nick, and Denis Alexander. 2009. *Rescuing Darwin: God and Evolution in Britain Today*. Theos.

Teixeira, Pedro, and Marcelo Andrade. 2014. "Entre as Crenças Pessoais e a Formação Acadêmica: Como Professores de Biologia Que Professam Fé Religiosa Ensinam Evolução?" *Ciência & Educação (Bauru)* 20 (2). Universidade Estadual Paulista Júlio de Mesquita Filho.

Tidon, Rosana, and Eli Vieira. 2009. "O Ensino Da Evolução Biológica: Um Desafio Para o Século XXI." *ComCiência*, no. 107. SciELO Brasil: 0.

Wiles, Jason R, and Brian Alters. 2011. "Effects of an Educational Experience Incorporating an Inventory of Factors Potentially Influencing Student Acceptance of Biological Evolution." *International Journal of Science Education* 33 (18). Taylor & Francis: 2559–85.

Chapter Two

Creationists or Evolutionists?

High School Students' Conceptions on the Origin and Evolution of Life

Pedro Teixeira

Evolution is acknowledged as a central axis for biology since it offers an epistemological framework to unify different areas of knowledge under the same discipline. As Dobzhansky (1973) says, nothing in biology makes sense except in light of evolution. For this reason, researchers and curriculum documents have argued that evolution should be one of the main guidelines of biology education (Brasil 2013; Sanders and Ngxola 2009). Several studies in Brazil and in the rest of the world indicate, however, that the teaching of evolution faces many difficulties. They are linked to textbook misrepresentations of Darwin's and Lamarck's theories, to teachers' and students misconceptions of the evolutionary theory and the nature of science, as well as the relationship with religious beliefs (Almeida and Falcão 2010; Smith 2010; BouJaoude et al. 2011; Young and Nehm 2011).

Since Darwin's times, some religious groups have demonstrated opposition and rejection to evolutionist ideas. By the end of the nineteenth century in the United States, conservative protestant groups defended reading the Bible in a more literalist way than the liberal ones. Following the publishing of the book series *The Fundamentals*, this supradenominational group, the Fundamentalists, organized a strong resistance movement against the teaching of evolution in schools. Many states approved anti-evolution laws. In Tennessee, this led to the notorious "monkey trial," when the high school biology teacher, John Scopes, was tried and convicted for teaching the then-prohibited theory. Later, these laws were revoked, but from then on numerous court disputes have taken place regarding the inclusion or exclusion of the evolution theory or creationism from the school curricula. In general, these attempts tried to promote creationism as a scientifically valid alternative to evolution, being repeatedly rejected in court. The latest major court decision, ruled that the so-called Intelligent Design is not scientific, but a modified version of creationism. As such, it cannot be taught in science classes.

In Brazil, the creationist movement and the anti-evolutionist initiatives are not as strong as in the United States, but as Dorvillé and Teixeira (2015) argue, they have four main ways of acting. Firstly, the two oldest still active creationist institutions are the Brazilian Creationist Society (founded by Adventists in 1972) and the Brazilian Association for Creation Research (founded in 1979 and linked to the Institution for the Creation Research, founded in the United States by the Young Earth creationist Henry Morris). These two institutions promote events on creationism that mimic scientific conferences, bringing national and foreigner speakers. Secondly, in 2014, the Brazilian Intelligent Design Society (TDI-Brazil) was created, led by the chemist Marcos Eberlin. In their manifesto, they defend a scientific creationism in opposition to a religious and philosophical one. Thirdly, these groups argue that teaching only evolution in schools is not fair. Instead, creationist views should also be taught, in favor of plurality of points of view and promoting the teaching of the controversy.

The fourth way is related with the rise of Pentecostal Evangelical politicians and the increasing Pentecostal believers in the country. Pentecostalism has its origins, by the end of nineteenth century, in the United States and the first Pentecostal churches in Brazil were founded by missionaries in the first decade of the 1900s. Brazilian Pentecostalism is comprised of denominations that emphasize the gifts of the Holy Spirit (such as healing and speaking in tongues). They also tend to perform literal readings of the Bible and to reject positions of the mundane (everything that is seen as not belonging to the religion is considered diabolic) (Mariano 1999; Mendonça 2006).

For the last thirty years, the Pentecostals are the fastest growing religious group in Brazil. In 1980 they made up 3.2 percent of the population and in 2010, 17.8 percent (IBGE 2012). They have grown in outskirts of large urban centers, especially Rio de Janeiro and São Paulo. Most believers are poorly educated and live in relative poverty. Researchers point out that the conservative rhetoric of these churches work as a way to overcome the precarious living conditions of these areas (regarding health assistance, education, unemployment, and drug addiction, for instance) (Freston 2013; Mariano 1999).

The Pentecostal denominations have shown a rising resistance toward evolution theory, especially through the publication of books and DVDs and by trying to pass legislative bills for introducing creationism in public and private schools (Dorvillé and Selles 2009; Malafaia 2009; Brasil 2014, 2016). Silas Malafaia, one of the most famous Brazilian Pentecostal pastors, says that scientists and teachers have a "*satanic interest* (. . .) [to] *blind the understanding of non-believers so that they have no knowledge of the truth and be set free by it*" (Malafaia 2009, 33). Marco Feliciano and Jefferson Campos, Pentecostal pastors and congressional representatives, presented bills to introduce creationism in schools curricula. They argue that it is not fair to teach only the evolution theory, since the majority of Brazilians are Christians and this would limit the students getting to know other beliefs. These bills have not yet been voted nor approved up to this date.

Previous studies regarding evolution teaching in Brazil show that Pentecostal and other Protestant students and teachers are more resistant to evolution ideas. Dorvillé and Selles (2009) interviewed Pentecostal pre-service Biology teachers and found out that most of them either reject the scientific approaches to the origin of living beings or try to conciliate them with their own religious beliefs. Similar results were found by El-Hani and Sepúlveda (2010) who interviewed five Protestant pre-service teachers (charismatic and mainstream Baptists). Oliveira and Bizzo (2015) investigated the acceptance and understanding of evolution of more than 2,400 high school students in the five regions of the country. Their results show that less Pentecostal students consider to be true items related to the age of the Earth, fossils, human evolution and evolution as a source of diversity than the students of other religions. Oliveira, Bizzo, and Pellegrini (2017), in a survey with Italian and Brazilian students, found that Pentecostal students in Brazil and non-Catholic Christians in Italy have a greater acceptance of divine planning and the creation of human beings, in comparison with their countrymen from other religions.

This work focuses on the acceptance of evolution by high school students in Rio de Janeiro, Brazil. More specifically, we wanted to know the students' conceptions about the origin of life, human evolution and how they are related to their religious beliefs. We surveyed students from two schools, one with more Pentecostal students than the other. Observations of biology classes and other spaces were conducted for nine months (from February to October in 2014) on a weekly basis and twenty students were interviewed (ten from each school). The results show that the majority of Pentecostal students express views close to the literal text of the Bible. However, students of other religions expressed similar views. Many of the interviewees try to conciliate their religious beliefs with parts of the scientific knowledge. Moreover, even students who had no religion believed in a teleological view for evolution or in some kind of agent that would drive it. We conclude that this diversity of thoughts challenge science educators to promote the understanding of the evolution theory.

METHODOLOGY

The Participant Schools

This research was conducted in two public high schools in Rio de Janeiro state, Brazil. One of them, Guarani,[1] is located in Gávea, an upper-middle-class neighborhood in the South Zone of the capital, but is majorly attended by low-income students. The other one, Passaredo, is in Nilópolis, a lower middle class city in the capital's outskirts. It is also attended by low-income pupils. We chose to investigate these two schools due to the distribution of the Pentecostals, according to data from the Demographic Census of 2010 (IBGE 2012). Nilópolis has a greater presence of Pentecostals than Rio's southern area. As such, we could compare two different contexts for similarities and differences regarding the students' religious profiles.

Guarani is a large school, with more than 2,400 students at the beginning of the 2014 school year, divided into three shifts (morning, afternoon and night) and 133 teachers. It is composed of a three-story building with twenty-one classrooms, library, reading room, computer laboratory, video room, yard, two sports courts, a large auditorium, a small cafeteria, and a parking lot. Additionally, there is a biology laboratory, but it was never used during our observation period. All classrooms have air conditioners and multimedia projectors. Guarani is located near a large hospital, a private university, a shopping mall, cinemas, theaters, and residential buildings.

Passaredo is smaller than Guarani, with about 1530 students and 108 teachers at the beginning of 2014. It has two buildings: a three-story one and another two-story one. In the first building, there are fourteen classrooms, a principal's office, secretariat and teachers' rooms as well as the cafeteria. In the second, there are two more classrooms, the library, and a science lab. Like in Guarani, the laboratory was never used during the time of observation of the research. There is also a multisport court used for games during recess and physical education classes. Passaredo is surrounded mainly by small houses, businesses, and many Pentecostal churches (more than fifteen on the four blocks closest to the school). Thus, the Pentecostal presence was more evident there than around Guarani.

The Participant Students

In 2014, Guarani School had twenty-six classes of first year of high school, the majority being in the afternoon shift, which we opted to study. Passaredo school had six classes of first year of high school, all in the morning shift. We observed ten classes in Guarani and four in Passaredo for nine months. We observed few conflicts involving religious beliefs and scientific knowledge. Both schools had approximately the same number of girls and boys, most students were non-white, and sixteen years old or younger. In Guarani, 37.8 percent were Catholic, 23.7 percent Protestant (15 percent Pentecostal) and 28.5 percent believed in God but had no religion. In Passaredo, 8.5 percent were Catholic, 52.5 percent Protestant (43.7 percent Pentecostal), and 25.5 percent believed in God, but had no religion.

From the observations, we chose to interview ten students. Most of them were participative of the classes, asking questions and showing interest in what was being taught. Others demonstrated involvement with religious activities, reading religious books or talking about it with friends. In Guarani, we interviewed three Pentecostal students (G1, G3, and G7), three Catholic (G2, G4, and G6), two without religion (G8 and G9), one Atheist (G5) and one who declared to be both Pentecostal and Catholic (G10). In Passaredo, we interviewed eight Pentecostal students (P1, P2, P4, P6, P7, P8, P9, and P10), one Catholic (P3) and one without religion (P5).

Research Instruments

The interviews were semi-structured. We followed a script previously designed, but we were open to new information and to probe as much as possible many of the answers. The questions explored students' religious beliefs, their relationship with their religion, and their beliefs concerning the origin of life and the origin of human beings:

1. Do you believe in God?
2. What is your religion?
3. How often do you go to Mass/services?
4. How important is your church to you?
5. Do you attend a group of your religious community (e.g., youth group)? Which one? Do you do any activity in your religious community? Which one?
6. Are your parents/legal guardians religious? What is their religion? Do they encourage you to participate in their religion? If you do not participate, what is their reaction?
7. Do you attend to religious education classes? Why?
8. How would you explain the origin of life on the Earth?
9. How do you think human beings appeared on Earth?
10. How do you feel when the teacher says that life may have arisen by natural processes, without the action of God?
11. Have you ever felt discriminated because of your religion in school?

We interviewed the students during their recess and the interviews were from twenty to thirty minutes long. The interviews took place in quiet places, such as a reading room, and none of the schools' personnel were present, in order to avoid students' embarrassment. A questionnaire was applied to all the students in the first year of high school in both schools. For this chapter we only use part of its results, concerning the religion of the students. A deeper analysis of it can be found in Teixeira (2019). The university's ethical committee approved the research, and consent terms were distributed to parents, legal guardians, and the students.

RESULTS AND DISCUSSION

Overall, eleven students gave statements close to a literal view of Genesis, one a view of day-age creationism, six tried to conciliate scientific and religious conceptions, and two manifested a view closer to a scientific approach. There are distinctions within these groups, but many times, it is difficult to see clear boundaries between these categories.

Most of the students interviewed said they believed in the literal account of Genesis. One Catholic (G2), two without religion (P5 and P9) and eight Pentecostals (P1, P2, P4, G7, P6, P7, P8, P9). These students generally indicated that they believed that God created the universe, the Earth, living beings, including humans, in seven days (24-hour intervals). G2's speech synthesizes these ideas:

> *Researcher (R):* In your opinion, how did life on our planet come about?
>
> *G2:* What a difficult question . . . [. . .] I believe in Adam and Eve, the capital sin. I believe . . .
>
> *R:* Do you believe in what?
>
> *G2:* That God made the Earth in seven days and in the last He rested. And that He made things. Made the light, made the Sun . . . I believe in this creation. I think the theory of creationism, that speaks of the creation of seven days, I believe a lot in it . . . Not of evolutionism, right?
>
> *R:* So you agree with what's in the Bible?
>
> *G2:* Yes.
>
> *R:* Even when it talks about seven days?
>
> *G2:* Of course. [. . .] I'm not sure about this history of the Big Bang . . . That it was there and exploded and then the Earth was made . . . Then the species that was in the water grew and . . . Ok . . . But how it grew up? Ok . . . I may

study, learn, that there were microorganisms. Maybe the sun temperature and everything that was . . . Ok, but how did the species that we know today was born from it? Of course, I do not recriminate people who believe in this, but I do not think it is plausible either.

In this excerpt, the student says that life and our planet arose by the act of God. It is important to note that, in his view, creationism is more plausible than scientific theories about the origin and evolution of life. He argues that even though he studies the conditions for the first living organisms to appear, he still believes that scientific theories cannot explain how they came about. It would not be possible, therefore, that there was no creative agent behind these processes. Moreover, the student refers to creationism as a theory. Following Evans et al. (2011), this view could be considered as two competing cultural models (science and supernatural) dominated by the supernatural. Thus, for these students, evolution and creationism would be located at opposite poles, with visions that would not be reconcilable.

Four students (P1, P2, P6, and P9, Pentecostals) reinforce the importance of faith and personal beliefs to accept creationism. P1 and G9 (without religion) argue that scientists reject religion, and, because of that, they do not accept the theory of evolution. In turn, P1 and P2 highlight the influence of family and friends for the construction of their conceptions. P1 brings together several of these ideas:

R: In some biology classes, the teacher talks about the theory of the origin of life. The teacher sometimes speaks of a theory that does not necessarily involve God's interference. What do you think of that? How do you feel when the teacher talks about it?

P1: [. . .] I still think God created the world. I grew up with it. But, I do not judge the scientists who think that, who don't have a religion, who are atheists. They want justifications. Because religion, evangelism is complicated, because you have to have faith to believe. You got to have faith, you have to feel. You got to have faith to believe. He made man from dust of the ground. He created man. Some scientists have found traces of dust in us, in our body, and they cannot explain it. My colleague told me. It's complicated, but it's a matter of faith.

Despite school classes, the student still believes that God created the world because that is something she grew up with. She does not judge scientists who accept scientific theories, suggesting that they have no religion and are atheists. Then, she argues that whereas the scientists want justifications one must have faith to believe the Bible. In this sense, because scientists would not have faith, they could not believe in creationism. Finally, she adds that, according to a friend of hers, researchers found traces of dust in humans, which would be a scientific evidence of God's action.

P1 interconnects several ideas with her religious belief to fulfil a double-function: (1) to support creationism and (2) to reject the scientific theories of origin and evolution of life. When she says that scientists have no religion, she suggests that science and religion are incompatible, for it would not be possible to be a scientist and religious at the same time. When she says that scientists want justifications and evangelism demands faith, it can be inferred that those who have faith in evangelism need no justification for their beliefs. Yet, she argues using an allegedly evidence of the natural world, that humans would have traces of dust in their bodies. It is also worth mentioning the importance given to her friend as a safe source of scientific research. P1 might think that the biology teacher would be a more appropriate resource for information on academic research. However, the knowledge discussed by the teacher contradicts her personal beliefs and/or the teacher would be one of those who has no faith. Thus, she privileges her friend, who implicitly seems to also be religious and helps strengthen her creationist conceptions.

Within this first group, there are still six students (P1, P2, G9, P6, P7, P8) who reject the neo-Darwinian theory for human evolution because it allegedly says that "humans came from monkeys." G9 and P8 said it is "funny" to think that humans could have come from monkeys and P7 said it was "weird." P1 and P2 argue that if humans came from monkeys, there would still be monkeys turning into humans, which does not happen. P6, on the other hand, compares scientist with the Egyptians, who worshiped other gods.

> *P1:* I do not believe in any of this monkey business. To make it clear. [Laughs] Yes, the monkey is very intelligent, very intelligent. But if man came from the monkey, would there be monkeys now? No, it wouldn't. For me, this theory should have been ended, in quotation marks, a long time ago, because it has no justification. How? To come from monkeys . . . Because if it came from monkey, the monkey was going to become a man through time and it did not happen.
>
> *R:* You said there are people who say that man comes from monkey and that is a lie. Explain why do you think this is a lie.
>
> *P6:* Because I think it's unlikely. [. . .] I do not have a concrete answer to this, but I think it is not true that the man came from the monkey and there are other people who say that . . . In Egypt's time, things like that . . . People worshiped other Gods, that the Pharaoh . . . Things like that . . . Images of Gods . . . And our God hadn't any image, and was criticized . . . People say that with evolution we are having now, people say that the human being came from monkey . . . For me, this is not right . . . For me, it is a lie [emphasizes]. People make up something that has nothing to do with it. The very truth is that God created us, made Adam and Eve and everything. The origin of life was God who created us.

The premise that "humans came from monkeys" is not correct from a scientific point of view, nor is it postulated by neo-Darwinian evolutionary theory. Nevertheless, the argument supported by P1 and P2, though motivated by a religious conception, is based on elements of the natural world derived from their own experiences (observation that monkeys are not turning into humans). On the other hand, the argument of P6 is based exclusively on his religious conception and not on evidences of the natural world. These different ways of resisting neo-Darwinian evolutionary theory were also pointed out by Thagard and Findlay (2010). There are cognitive obstacles (conceptual, methodological, and coherence difficulties) and emotional obstacles (linked to psychological and/or religious factors). In the case of P1 and P2, the resistance is due to conceptual difficulties of the evolutionary process and understanding the scientific methodology. In the case of P6, the resistance is related to an emotional and religious factor, considering that it rejects the scientific vision for the origin of human beings because this one deviates from the established truth of God.

Three students (G1, G3, and P10, Pentecostals) mix religious and scientific elements, but they are uncertain of how they are articulated. They say they tend to believe the Bible, but also believe that certain scientific knowledge about the origins makes sense:

R: How do you think life on our planet came about?

G1: The Bible say there were Adam and Eve. I do not know. First, He put . . . Let there be fire, then let there be Earth . . . Let there be light . . . And then started creating Earth. But in science it says that it was through the Big Bang that it came about . . . That the Sun exploded, and such . . . But I think it was through God and science helped.

R: How do you think humans came about on our planet?

G1: Through Adam and Eve [laughs slightly].

R: Why did you laugh when you spoke about Adam and Eve?

G1: Because people do not believe. They say it was through the Homo sapiens sapiens, right?

R: And what do you think when people say that?

G1: I think it was . . . We were not there to know if Adam and Eve is really like man was formed . . . No! What if Adam and Eve was a Homo sapiens sapiens that was evolving and that little by little came to what we are now. By an evolution.

G1 answers the first question by saying that the Bible speaks of Adam and Eve and of the stages of divine creation. Then she speaks of the scientific view, referring to the Big Bang, understood as the explosion of the Sun. Soon after, she says that God created the world, but science has "helped." In the following questions, when discussing the existence of Adam and Eve, the student says that it is possible that they really existed but might also have been part of the evolutionary process of human beings.

It is possible to see that G1 recognizes the validity of certain elements of religious and scientific thought. We can see that when she says that she believes that life came "through God and science helped," although science itself only seeks to study the conditions for the origin of life and has not acted directly in this process. Secondly, the student points out that no one was there to see if Adam and Eve did exist. The student emphasizes direct observation as a way of constructing definite knowledge about a certain phenomenon. However, she uses such an argument to raise doubts about the existence of Adam and Eve, something that reinforces when considering the possibility that they were part of human evolution.

We can understand these students' conceptions as an example of juxtaposition (Evans, Legare, and Rosengren 2011). That is, a form of coexistence of scientific and supernatural beliefs in which both are used to explain the same phenomena in a non-necessarily organized way. Even if they show some doubts, for them evolution and creationism would not necessarily be rivals, as defended by the previous group.

A student (G10, Pentecostal and Catholic) expressed in his interview a position closer to day-age creationism (Scott 2005):

> *R:* How do you feel when a teacher talks about the origin of life and evolution, and talks about explanations that do not involve God?
>
> *G10:* I, myself, in my own head, I try to fit it in with what the Bible says.
>
> *R:* What do you mean? What do you get from one and what do you get from another to fit in?
>
> *G10:* For example, the question of time. As for God, it took seven days to create, all the formation of the Earth for us might have been millions, millions of years. And what God did like that [snapped his fingers], but to us, it was a slow thing in our view, a slow formation. Let's say that our evolution, human evolution, was like . . . For God, he was forming us, while from our point of view, we were gradually evolving. I fit things in like that.

In this passage, G10 says that he tries, on his own account, to use elements of religious and scientific knowledge to construct his personal vision on the origin and evolution of life. He does not rule out the existence and activity of God in these processes. Yet, he points out that the days of creation were intervals of millions of years rather than twenty-four hours, for the perception of time for God is different from the human being.

Three students (G4 and P3, Catholics, and G5, atheist) expressed visions closer to theistic evolutionism (Scott 2005). They do not accept that the universe could have arisen "out of nowhere" and attribute its emergence to a force (in the case of G5) or God (in the case of G4 and P3). As for the origin of life, G5 believes that the same force also acted in this process, while G4 and P3 say they believe that only natural phenomena would be involved:

R: How do you think life began on our planet?

G5: At this time, I say I'm a bit agnostic. [. . .] Let's give an example. A meteor fell and this meteor had some kind of radioactivity and began to appear some kind of living thing. So how did this meteor appear? Out of the blue? Everything came out of nowhere? So somehow, I believe there is something. There is some sparkle there that has the strength of something, but today I do not call it God. [. . .] In other words, what I believe is that there is some greater force, but I do not call it God. I do not call it anything, and I do not believe it. I just know it must exist. It certainly exists, but to this day I do not know what it is. As there are many people who really do not know it, maybe everyone does not know.

R: How do you think life began on our planet?

G4: On our planet, I think it really was the way we have proof. What science explains. Of evolution and such. I do not believe that Catholic God created the man from dust of the ground and created the woman from the man's rib, I do not believe it. But I believe God, the God that I believe . . . I will not know how to differentiate . . . The God that I believe, he created the universe and the universe that he created propitiated the essence of life on some random planet and it turned out to be Earth.

Student G5, while initially declaring himself an agnostic, claims that a force acted at the origin of life on Earth. He cannot define it accurately and says he does not believe it. While it may be incoherent to claim that something exists and at the same time to say that one does not believe it, the student's speech seems to indicate that he has no religious faith in that force. G4, in his turn, says that life has arisen on our planet as scientific theories describe, rejecting a literal reading of Genesis and Adam and Eve. Differentiating his notion of God from that which would be more common to Catholics, he believes that this transcendental agent created the universe with characteristics that would favor the emergence of life on some planet, that happened to be ours.

Still in relation to this subgroup of students, P3 considers possible that intelligent extra-terrestrial beings have acted in the emergence of life on Earth:

R: When the teacher says that life may have come up without the interference of God, what do you think of it? How do you feel?

P3: [. . .] I think it might have been aliens, in quotes. After watching these movies, *Prometheus* and all. It may have come through the hand of a being, something greater that we might call aliens or ETs. I think by the fact that people say they're an intelligent race, I think they got the gift of giving life . . . Like this *Prometheus* movie . . . The molecule of one of them dead, when a lightning bolt struck the tree and the human appeared. If God did not do that, I think it was the aliens.

Of all the students interviewed, P3 was the only one to raise the possibility of intelligent beings from other planets acting on the origin of life on Earth. Citing the film *Prometheus* as an influence for his conception, the pupil says that these aliens have managed to obtain the capacity to provide life because they were intelligent.

It is important to note that for these three students, certain natural processes involved in the origin and evolution of life may occur without the performance of a creator. However, when they cannot find a satisfactory explanation based on scientific knowledge, they resort to the idea that there is a creative agent. As discussed by Blancke et al. (2011), this agent is not necessarily a figure based on a religious view (as G5 refers to a force and P3 to extraterrestrials). The authors argue that this is an intuitive factor and a possible obstacle to the teaching of evolution.

Although they have different characteristics between them, it is possible to find similarities between the thought of G4, G5, P3, and G10. These four students articulate scientific and supernatural knowledge in a more organized way than the previous group and, mainly, they advocate a complementarity between one and the other. In view of the debates on science and the supernatural, Evans et al. (2011) call *parallel reasoning* this kind of coexistence of epistemologies, when two or more are used to explain similar or equal results.

The other two students, G6 and G8, accept scientific theories for the origin and evolution of life on Earth. G6 declared himself Catholic and when asked how he thought life had begun on our planet exposed two visions, one closer to the Bible and another to the scientific theories and said he believed in both. Next, however, he stated that he tends to believe more in the conception of science:

R: In some biology classes, the teacher talks about theories of the origin of life, how life arose on our planet and some of these theories say that life may have arisen without the intervention of God. What do you think? How do you feel when the teacher talks about it?

A: In my opinion, neither is wrong. If you look on the side of science, it possibly may not have been God, but if it is on the side of God, God created the Earth.

R: What is your opinion about how life on our planet came about?

A: I believe that the Big Bang happened. [. . .] The universe expanded, as time passed, the heat of the Big Bang was forming the Earth, then the Earth . . . The Earth began to rise, vegetation, animals began to appear . . . Then monkeys, as we are descended from monkeys, I can say that, they were evolving and getting to us. In the other part, as in God's part, I can say this: God created the universe, created the Earth, took seven days to create the Earth. Then created a human being and took out a rib and created the woman.

R: Which one is closest to your vision?

A: Science. More in science, actually.

R: Why? Why do you believe in this view of science more than in the religious view?

A: Because, every scientist says, the Big Bang was an energy that eventually expanded the universe and created the planets. [. . .] I believe in science because, like, the way they talk is very realistic, much more real than the story of God. I believe more in science. [. . .] Because they've done a lot of research to come to that conclusion.

It is worth noting that the student first states that he does not believe either side is incorrect, which points to a reconciliation of both. However, when we question him more directly as to how he could believe "on both sides," the student only explains how he understands scientific theories and religious knowledge. He does not make connections between them, as other students did by referring to God acting through processes that science investigates. After we asked him a third time which one he believes most, the student says that it is the scientific thought, because it would be more real. As we asked why he found it more real, G6 refers to the authority of scientists and the time they devoted to studying the processes of origin and evolution of life. It should be noted, however, that this argument also seems problematic, since it does not take into account the large number of empirical evidence and the logic of scientific research. Although the student seems to accept evolutionary theory, his understanding could still be deeper and based on more reflections.

The student shows certain similarity with the previous group, since he expresses, up to a point, to accept both the scientific and the religious view. However, we consider it different from those, because at the end of his speech he indicates that it would be more susceptible to accept the scientific epistemology. Thus, he points out to two competing cultural models (science and supernatural), with predominance of science (Evans, Legare, and Rosengren 2011).

On the other hand, G8 said not to believe in God, but affirmed to believe in reincarnation and in a force of nature. This force would not be a creative agent but would act to make people rewarded positively or negatively for their attitudes. Regarding the origin of life, the student expresses a view close to scientific thought:

R: Thinking about the origin of life, how would you explain the origin of life on our planet? How did life on Earth come about?

G8: Well, by evolution, right? Like the biology teacher says. I believe in it. I do not believe in Adam and Eve. Because there are so many ethnic groups, Asian, black, Indian. And there is no way that two people can give life to so many ethnicities, since they are the same. So, I believe that people are evolving, transforming. Mutation as well.

R: But what do you think . . . When the teacher says that life can arise without the interference of God, in a force . . .

G8: I believe in it. Because it's an easier thing to believe.

R: What relationship do you see between the origin of life, evolution, and the force you said you believe?

G8: No. It has nothing to do with it. I guess not. This force I'm talking about is in the human being, you know? It is not in the head of the animals that were there. If there was a force it was because of the universe, I do not know.

R: To generate the universe, this force . . . How would that force have acted on it?

G8: I do not know, nature, right? The force of nature . . . I cannot explain, but I think that's it.

P: Your very opinion . . . How do you think about it?

G8: I do not know . . . It has been so long . . . The little things can evolve . . . Bacteria that live in the sea, they may leave the sea. . . . I do not know, I have a lot of doubts, you know . . . I do not have a fully formed opinion. [. . .] One day, I saw a documentary about the mermaids that exist, but not the way we think . . . Then you see that it's all a lie . . . Wow . . . You cannot believe everything you see. That's why I do not believe the Bible. You cannot believe everything you read.

G8 believes that life arose according to scientific theories that she studied in her biology classes. She argues that she does not believe in Adam and Eve because it would not be possible for several ethnicities to be formed from two equal individuals. She adds that humans go through evolutionary processes and mutate. When asked about the relationship between the force in which she believes and the origin of living beings, she says there is no connection. The student explains that this force would be inherent to humans and not to other animals. In addition, the force that would be responsible for the origin of the universe and life would be a "force of nature," which she cannot define precisely. Asked for more details, G8 said she was not entirely sure and referred to a documentary she watched and whose truthfulness she questioned. She further adds that this is why she does not believe in the Bible, for one should not believe everything one reads.

The student accepts scientific theories for the origin and evolution of life and seeks empirical explanations to justify her opinion, such as the diversity of ethnicities of human beings. Thus, she not only assumes the position of scientists as true, like G6, but demonstrates certain reflection to develop her vision. Secondly, there are two definitions of "force" used. The first, restricted to humans, resembles a person's motivation to accomplish something. The second is related to nature as a whole, indicating a certain intention about its processes. As pointed out by Blancke et al. (2011) and Evans et al. (2011), teleology is an intuitive idea of humans about the processes around them and that is as an obstacle to the understanding of neo-Darwinian evolutionary theory. In this sense, it is important to note that even a student who seems to accept scientific theories for the origin and evolution of life, presents ideas that may limit their understanding.

In order to clarify and summarize the positions of the students about the origin and evolution of life, and more specifically, human beings, we prepared the two continua of figure 2.1. The closer to the left, the more literalistic and

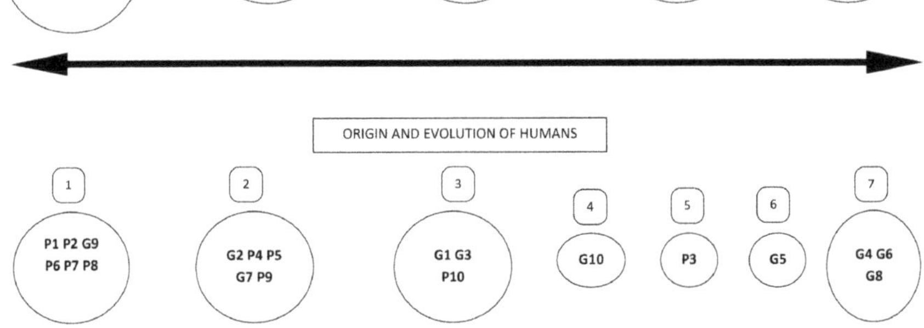

Figure 2.1. Interviewed students according to the continua of conceptions about the origin of Earth, life and human beings. Pentecostals: G1, G3, G7, P1, P2, P4, P6, P7, P8, P9 and P10; Catholics: G2, G4, G6 and P3; Atheist: G5; No religion but believe in God or in a Force: G8, G9 and P5; Pentecostal and Catholic: G10.
Source: The author. Created by Pedro Teixeira.

more resistant to evolutionary theory the students are. The closer to the right, the more the students accept the theory of evolution.

In the upper continuum, there are five distinct groups, which have been described above. The first, "Literal," is that of students who adhere to a vision close to the literal text of Genesis and is formed by G2 (Catholic), G9 and P5 (without religion), P1, P2, P4, G7, P6, P7, P8, and P9 (Pentecostals). For these students, God created life, Adam and Eve existed, and the days of creation were twenty-four-hour intervals. The second, "Juxtaposition," brings together learners who express doubts about certain passages in the Bible and accept certain scientific elements, although they do so in a confused way. This group includes students G1, G3, and P10 (Pentecostals). The third group, "Day-age," is formed by only one student, G10, Catholic and Pentecostal. He believes the days of creation would be intervals of millions of years, which would allow him to reconcile his religious beliefs with scientific knowledge about the evolution of living beings. The fourth group, "Theistic Evolutionism," encompasses students G4, P3 (Catholic), and G5 (atheist). They believe in neo-Darwinian evolutionary theory but have resistance in accepting scientific explanations for the origin of the universe and life. They believe that an agent, God, in the case of the Catholics, and a Force, in the case of the atheist, acted in these processes. Finally, the last set ("Evolution"), formed by G6 (Catholic) and G8 (without religion), groups students who demonstrate accepting scientific theories. It should be noted, however, that student G6 manifests an oscillating position between the religious and scientific view.

In the lower continuum, there are seven groups ordered according to the acceptance of neo-Darwinian evolutionary theory for humans. Since the dispersion and complexity of this point are greater than the previous one, the categories used in the works cited throughout this text seems to us not to capture them satisfactorily. Thus, we divided the interviewed students into seven different groups.

In the first set are students P1, P2, G9, P6, P7, and P8. These are the ones that express greater rejection to the evolutionary theory. They believe that God acted directly in the origin of human beings, creating Adam from dust and Eve from one of his ribs. In addition, they say they do not believe in evolution because it says that humans came from monkeys, which would not be true, since we still find monkeys today and these are not turning into those.

The second group, formed by the students G2, P4, P5, G7, and P9, has the same characteristics of the first, except for the reference to monkeys. The third set involves the students G1, G3, and P10, who express a vision that mixes certain scientific concepts with the belief that God had some participation in the origin of human beings. The fourth is restricted to student G10, who believes in day-age creationism and that human evolution is the fruit of God's activity. The fifth is formed only by P3, who believes in the scientific theories for human evolution, but also thinks that aliens influenced this process. The sixth, in turn, encompasses G5 and G6, who accept human evolution, yet demonstrate some uncertainty. Finally, the seventh set comprises G4 and G8, for whom the human being was formed exclusively by natural processes.

We can see that the Pentecostals students tend to express literalistic views for the origin of life and human beings. Nevertheless, some Catholics and students without religion also have such ideas. We could note differences between the students from different religions, but we did not find differences linked to the school context. Thus, it seems that the students' religious beliefs are more important than a possible influence of a stronger religious environment around the schools.

It is noteworthy that there is a variety of interpretations and ways to conciliate religious and scientific concepts. Many of them involve teleological ideas and reference to an external agent, such as God, aliens, or a Force. As previous studies have shown, these might turn into obstacles to understand the theory of evolution (Blancke et al. 2011). Science educators should be aware of them, in order to explore different pedagogical strategies to avoid misconceptions. Still, as other researchers argue, it is important to avoid scientism as well. Attitudes such as to ridicule or belittle students' beliefs are not only ethically wrong, but also provokes what Cobern (1996) call cognitive apartheid. The student will wall off his or her beliefs, protecting them from a possible threat posed by the theory of evolution.

CONCLUSION

This study provides knowledge that might help in the reflection about evolution teaching. More specifically, it calls attention to the importance of the religious beliefs of the students in the shaping of their ideas, and possible constraints that might hinder the understanding of the evolutionary theory.

Certain interpretations of biblical texts might be obstacles to understanding evolutionary theory. However, it is important to note the position of different authors who have argued that the religiosity of students must be respected and that the goal of science education should be understanding it and not belief change (Cobern 2000; Reiss 2009; El-Hani and Mortimer 2007). Thus, the teaching of evolution should be oriented to its understanding as a scientific theory, constructed from empirical and multiple studies of evidence, without being imposed as an absolute truth to the student whose beliefs, perhaps, might clash with the vision of science. Far from exhausting the issue, we expect that this work might work as a starting point for new research hypotheses and for new strategic pedagogical approaches.

NOTE

1. The school and participant names are pseudonyms.

REFERENCES

Almeida, Argus Vasconcelos De, and Jorge Tarcísio Da Rocha Falcão. 2010. "As Teorias de Lamarck e Darwin Nos Livros Didáticos de Biologia No Brasil" [Darwin and Lamarck research programs historic-conceptual structure and its conversion to the school environment]. *Ciência & Educação (Bauru)* 16 (3): 649–65. https://doi.org/10.1590/S1516-73132010000300010.

Blancke, Stefaan, Johan Smedt, Helen Cruz, Maarten Boudry, and Johan Braeckman. 2011. "The Implications of the Cognitive Sciences for the Relation between Religion and Science Education: The Case of Evolutionary Theory." *Science and Education* 21 (8): 1167–84. https://doi.org/10.1007/s11191-011-9402-z.

BouJaoude, Saouma, Anila Asghar, Jason R. Wiles, Lama Jaber, Diana Sarieddine, and Brian Alters. 2011. "Biology Professors' and Teachers' Positions Regarding Biological Evolution and Evolution Education in a Middle Eastern Society." *International Journal of Science Education* 33 (7): 979–1000. https://doi.org/10.1080/09500693.2010.489124.

Brasil. 2013. *Diretrizes Curriculares Nacionais Da Educação Básica* [Basic Education Curricular Guidelines]. Brasília: MEC, SEB, DICEI.

———. 2014. *Projeto de Lei No. 8099: Ficam Inseridos Na Grade Curricular Das Redes Pública e Privada de Ensino, Conteúdos Sobre Criacionismo* [Law Project N.8099: Contents on Creationism to be included in curricula of public and private education]. Brasília: Congresso Nacional.

———. 2016. *Projeto de Lei 5336/2016: incluir a "Teoria da Criação" na base curricular do Ensino Fundamental e Médio* [Law Project N.5336: includes the Creation Theory in the curricular base of Elementary, Middle and High School]. Brasília: Congresso Nacional. http://www.camara.gov.br/proposicoesWeb/prop_mostrarintegra?codteor=1459211&filename=PL+5336/2016.

Cobern, William W. 1996. "Worldview Theory and Conceptual Change in Science Education." *Science Education* 80 (5):579–610. https://doi.org/10.1002/(SICI)1098-237X(199609)80:5<579::AID-SCE5>3.0.CO;2-8.

———. 2000. "The Nature of Science and the Role of Knowledge and Belief." *Science and Education* 9 (3): 219–46. http://libezp.lib.lsu.edu/login?url=http://search.ebscohost.com/login.aspx?direct=true&db=eric&AN=EJ605620&site=eds-live&scope=site.

Dobzhansky, T. 1973. "Nothing in Biology Makes Sense except in Light of Evolution." *American Biology Teacher* 35 (3): 125–29.

Dorvillé, Luís Fernando, and Sandra Escovedo Selles. 2009. "Conflitos e Tensões Entre Ciência e Religião Nas Visões de Mundo de Alunos Evangélicos de Uma Licenciatura Em Ciências Biológicas" [Religion, school and Science: conflicts and tensions in the worldviews of Biological Sciences students]. In *VIII Congreso Internacioanl Sobre Investigación En La Didáctica de Las Ciencias*, 2757–61. Barcelona.

Dorvillé, Luís Fernando, and Pedro Teixeira. 2015. "O Crescimento Do Criacionismo No Brasil: Principais Influências e Avanços Recentes" [The Growth of Creationism in Brazil: Major Influences and Recent Developments]. In *X Encontro Nacional de Pesquisa Em Educação Em Ciências*, 24–27. Águas de Lindóia: ABRAPEC.

El-Hani, Charbel Niño, and Eduardo Fleury Mortimer. 2007. "Multicultural Education, Pragmatism, and the Goals of Science Teaching." *Cultural Studies of Science Education* 2 (3):657–702. https://doi.org/10.1007/s11422-007-9064-y.

El-Hani, Charbel Niño, and Claudia Sepúlveda. 2010. "The Relationship between Science and Religion in the Education of Protestant Biology Preservice Teachers in a Brazilian University." *Cultural Studies of Science Education* 5 (1): 103–25. https://doi.org/10.1007/s11422-009-9212-7.

Evans, E. Margaret, Cristine H. Legare, and Karl S. Rosengren. 2011. "Engaging Multiple Epistemologies: Implications for Science Education." In *Epistemology and Science Education: Understanding the Evolution vs. Intelligent Design Controversy*, edited by M. Ferrari and R. Taylor, 111–39. New York: Routledge. https://doi.org/10.4324/9780203839638.

Freston, Paul. 2013. "Pentecostals and Politics in Latin America: Compromise or Prophetic Witness?" In *Spirit and Power: The Growth and Global Impact of Pentecostalism*, edited by Donald E. Miller, Kimon H. Sargeant, and Richard Flory, 101–18. New York: Oxford University Press.

IBGE [Brazilian Institute of Geography and Statistics]. 2012. *Censo Demográfico 2010: Características Gerais Da População, Religião e Pessoas Com Deficiência* [Demographic Census 2010: General characteristics of population, religion, and disabled people]. Rio de Janeiro: IBGE.

Malafaia, Silas. 2009. *Criação x Evolução: Quem Está Com a Verdade?* [Creation x Evolution: who has the truth?]. Rio de Janeiro: Editora Central Gospel.

Mariano, Ricardo. 1999. *Neopentecostais: Sociologia Do Novo Pentecostalismo No Brasil*. [Neopentecostals: Sociology of the New Pentecostalism in Brazil]. São paulo: Loyola.

Mendonça, Antonio Gouvêa. 2006. "Evangélicos e Pentecostais: Um Campo Religioso Em Ebulição" [Evangelicals and Pentecostals: A religious field in ebullition]. In *As Religiões No Brasil: Continuidades e Rupturas* [Religions in Brazil: Continuities and Ruptures], edited by Faustino Teixeira and Renata Menezes, 89–110. Petrópolis: Vozes.

Oliveira, Graciela da Silva, and Nelio Bizzo. 2015. "Evolução biológica e os estudantes brasileiros: conhecimento e aceitação" [Biological Evolution and Brazilian students: knowledge and acceptance]. *Investigações Em Ensino de Ciências* 20 (2):161. https://doi.org/10.22600/1518-8795.ienci2016v20n2p161.

Oliveira, Graciela Silva, Nelio Bizzo, and Giuseppe Pellegrini. 2017. "Evolução Humana e Religião: Opiniões de Jovens Brasileiros e Italianos" Human Evolution and Religion: Brazilian and Italian youngsters'opinions]. *Revista Brasileira de Pesquisa Em Educação Em Ciências* 17 (1):135–56. https://doi.org/10.28976/1984-2686rbpec2017171135.

Reiss, Michael J. 2009. "The Relationship between Evolutionary Biology and Religion." *Evolution* 63 (7):1934–41. https://doi.org/10.1111/j.1558-5646.2009.00714.x.

Sanders, Martie, and Nonyameko Ngxola. 2009. "Addressing Teachers' Concerns about Teaching Evolution." *Journal of Biological Education* 43 (3): 121–28. https://doi.org/10.1080/00219266.2009.9656166.

Scott, Eugenie C. 2005. *Evolution vs. Creationism: An Introduction.* Berkeley, CA: University of California Press.

Smith, Mike U. 2010. "Current Status of Research in Teaching and Learning Evolution: II. Pedagogical Issues." *Science and Education.* Vol. 19. https://doi.org/10.1007/s11191-009-9216-4.

Teixeira, Pedro. 2019. "Acceptance of the Theory of Evolution by High School Students in Rio de Janeiro, Brazil: Scientific Aspects of Evolution and the Biblical Narrative." *International Journal of Science Education* 41 (4). Routledge: 546–66. https://doi.org/10.1080/09500693.2019.1569283.

Thagard, Paul, and Scott Findlay. 2010. "Getting to Darwin: Obstacles to Accepting Evolution by Natural Selection." *Science and Education* 19 (6–8): 625–36. https://doi.org/10.1007/s11191-009-9204-8.

Young, Sun, and Ross H Nehm. 2011. "A Cross-Cultural Comparison of Korean and American Science Teachers' Views of Evolution and the Nature of Science" 33 (2): 197–227. https://doi.org/10.1080/09500690903563819.

Part II

EVOLUTION CURRICULUM AND INSTRUCTION IN BRAZIL

Chapter Three

Brazilian High School Biology Teachers' Perception of Evolution and Its Teaching

Heslley Machado Silva
and Eduardo Fleury Mortimer

The clash between religion and biological evolution has resurged in several parts of the world, notably in Brazil. To investigate how Brazilian high school teachers of biology conceive of evolution, a European questionnaire, BIOHEAD-CITIZEN, was applied to sixty-two teachers and seven questions were selected. Barbour's categories and the teachers' religion were considered in the analysis of the answers. Semi-structured interviews were also conducted with ten teachers and the discourse of the collective subject was analyzed. The questionnaire results showed that Brazilian teachers have difficulty in dealing with evolution and that they do not consider science and religion as independent fields. The data revealed that teachers attempt to dialogue with and integrate the fields to avoid conflict. Also, teachers' limited knowledge of biological evolution, and a complex and difficult arrangement to accommodate religion were found. The teachers' discourse revealed two obstacles in teaching evolution: their personal beliefs and the clash with the students' religiosity. Further study is needed to understand how these findings affect the teaching of evolution in Brazilian high schools.

New attempts to introduce creationism and Intelligent Design, its "Siamese twin," in the school curriculum have been made in various parts of the world, posing problems to formal scientific education (Silva 2017; Graebsch 2006). This initiative, which was previously restricted to the United States, has reached Latin America and grown especially in Brazil (Cornish-Bowden and Cárdenas 2007). We point out that this problem emerges in an environment of change in the religious profile of Brazilians. The number of Evangelicals has markedly increased, with some denominations being greatly refractory to the teaching of biological evolution. In contrast, the percentage of Catholics has decreased noticeably, while the number of self-declared non-religious people, including agnostics and atheists, has increased slightly (Mariano 2013; Silva and Mortimer 2014). In this context, several pastors, especially from the Neo-Pentecostal denomination, try to prove the evolutionary theory wrong by opposing it to any religious belief, increasing the potential for conflict (Malafaia 2009).

Simultaneously in Brazil, there has been an effort to introduce the teaching of creationism in the educational system and disseminate the pro–Intelligent Design movement (Abrantes and Almeida 2006; Sepulveda 2004; Silva et al. 2016), despite opposition from the Ministry of Education (Tidon and Vieira 2009). If, on the one hand, many of these new Evangelicals are refractory to any subject that might hurt their sacred scriptures, on the other hand, new atheists perceive Evangelicals as being backward and closed to contemporary science. To add up to this environment, 2009 was celebrated as the year of Darwin to mark his relevance (Ayala and Avise 2009), and reinforce the already strongly pointed central role of evolution in biology and science as a whole (Dobzhansky 1973; Valença and Falcão 2012). One way to better understand this movement is to investigate how biology teachers conceive biological evolution, how they deal with it in teaching and the difficulties they have. This context is very likely to affect the teachers and their perception of evolution, as well as their performance in the classroom (Sepulveda and El-Hani 2004).

THEORETICAL FRAMEWORK

The conceptions of the theory of evolution of the biology teachers interviewed were analyzed based on the following categories of relationship between science and religion: Conflict, Independence, Dialogue, and Integration (Barbour 1990). In the *Conflict* category, religion and science would tend to be mutually exclusive, irreconcilable, and definitely separate, each certain of its own truth. Radical atheist scientists and dogmatic biblical creationists are examples of those in this category. Barbour (1990) points out mistakes in this conception, since this category would not tolerate those who believe in the evolutionary theory and God, creating a hostile impasse to scientific education, especially in environments with a strong religious appeal. For the advocates of *Independence*, there is a clear division between science and religion, each being a distinct field of knowledge, with traditions that do not contribute to each other. Their themes, language, and methods are entirely distinct, non-competing, non-overlapping, and are mutually respected by their followers. Despite Barbour's (1990) observation that this category hampers the interaction between the two fields, he himself acknowledged that it avoids conflict. In this investigation of teachers' perceptions of evolution, this category was considered the closest to the ideal in an environment marked by dissension, such as Brazil. This view is supported by the concept of NOMA (non-overlapping Magisteria), which allows the mutual respect of differences and enables biology teachers to maintain both their beliefs and the scientific pillars of Darwinian evolution (Gould 1997a). Another category, *Dialogue*, points to a less rigid border between science and religion and collaboration between fields. Here, religious arguments can be used to fill the gaps on the origin of the universe not elucidated by science. We are cautious of this category in this study because teachers could make up for the flaws in their science training with their beliefs and preach them in classrooms. Finally, the *Integration* category, which has the faintest limits between science and religion, proposes a "natural theology" that would prove the existence of God through scientific logic itself. Those who support the Intelligent Design hypothesis fall into this category, because they believe that only through resorting to the participation of a creator is it possible to explain the complexity of organisms (Behe 1996). This contributes to making this category attractive to teachers who have limited scientific knowledge of biological evolution. However, this category entails its own risks, since it makes a pseudoscience hybrid of scientific and religious concepts possible.

METHODOLOGY

Quantitative and qualitative approaches have been used complementarily. Quantitatively, the European questionnaire BIOHEAD-CITIZEN (Carvalho and Clément 2007) was applied to sixty-two biology teachers from metropolitan Belo Horizonte, state of Minas Gerais, Brazil. The questionnaire had four-response option questions, sometimes using the Likert scale, also with four options. The teachers were selected based on accessibility and inclusion criteria: they should teach at high school and have worked on the biological evolution theme. From the 144 questions in the original BIOHEAD-CITIZEN questionnaire related to biology in general, seven that dealt with biological evolution were selected for this discussion. Of these, four were related to human origin and evolution (A.33, A.44, A.62, and B.28): 1) The emergence of the human species (*Homo sapiens*) was just as improbable as the emergence of any other species (A.33); 2) The purpose of the evolution of living species was the appearance of the human species (*Homo sapiens*) (A.44); 3) Which terms are more closely related to the origin of the human being: Adam and Eve, *Australopithecus*, Creation, Evolution, God, Natural Selection (A.62); and 4) Which of the following four statements related to the origin of humankind do you agree with mostly? Evolutionary process, Evolutionary process without God, Evolutionary process guided by God, God. (B.28).

Three questions are related to the acceptance of creationism and evolutionism (B.29a, B.29b and B.48): 1) Does the theory of evolution contradict my own beliefs? (B.29a); 2) Does creationism contradict my own beliefs? (B.29b); and, 3) The degree of importance of God in the evolution of the species (B.48). The responses were analyzed based on the categories proposed by Barbour (1990) and the teachers' religions. In the qualitative approach, ten biology teachers were interviewed. They were chosen based on their availability and the same inclusion criteria used for the application of the questionnaire.

A semi-structured interview was developed that allowed the teachers to freely present their vision on the teaching of biological evolution, their conflicts and difficulties. The interviews were analyzed through the Collective Subject Discourse (DSC) methodology, which seeks to reveal the subjects' social representations while preserving the individuals' perception (Lefevre and Cavalcanti Lefevre 2014) at the same time. Following this method, individual opinions with similar meanings were grouped into general categories based on the answers given to the interview questions, identifying the meanings expressed, and the main and latent ideas. Finally, we synthesized the key expressions that presented ideas on the teaching of evolution in first-person discourse.

KEY FINDINGS

The percentages of types of religions practiced in the sample of biology teachers investigated was very different from that in the Brazilian population in general, especially Evangelicals—11 percent in the sample and 22.2 percent in the Brazilian population (Census 2010)—and Spiritualists, with 24 percent in the study sample, and 2 percent in the Brazilian population. The individuality of the sample—a group of biology teachers from a metropolitan region of Brazil—accounts for the differences, which must be born in mind. Figure 3.1 shows the percentages of types of religions of the teachers interviewed in comparison to the general Brazilian population.

The results show a wide variation in the biology teachers' responses regarding Barbour's categories. A small number of responses indicated Independence between the fields of science and religion. The Conflict, Dialogue, and Integration categories prevailed.

Each question was analyzed under the light of the percent results of the responses. The results of the objective questions were crossed with the teachers' discourse on the themes of biological evolution and the origin of human beings.

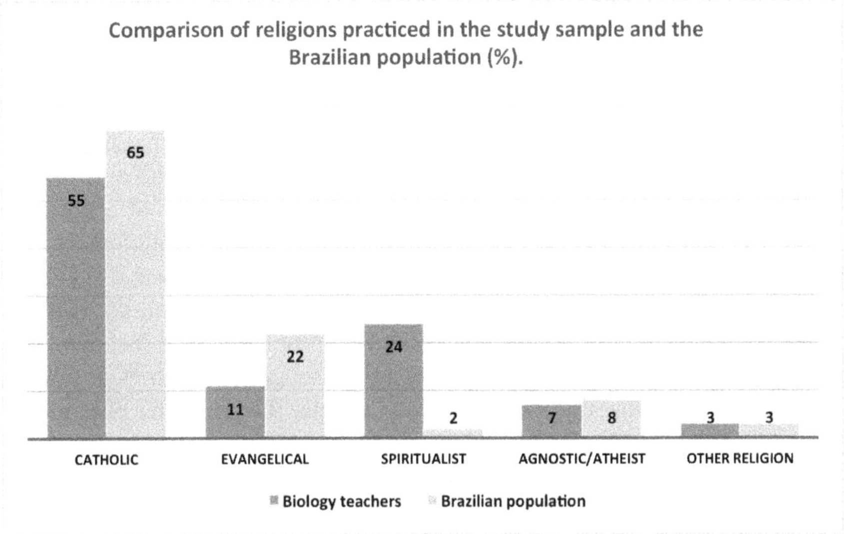

Figure 3.1. Comparison of religions among the teachers investigated (research data) and the Brazilian population in general
Data taken from Mariano (2013), graph created by Heslley Machado Silva and Eduardo Fleury Mortimer.

Was the Origin of Human Beings a Natural Phenomenon Like That of Any Other Animal Species?

The results of Question A.33, "The emergence of the human species (*Homo sapiens*) was just as improbable as the emergence of any other species," were: 39 percent of the teachers totally agreed; 13 percent agreed more than disagreed; 12 percent disagreed more than agreed; and 36 percent totally disagreed.

Although a great percentage of the Brazilian teachers interviewed agreed that the origin of human beings was as improbable as that of any other species, it stood out that nearly 50 percent of the biology teachers do not agree with this statement. We may ask why professionals graduated in biological sciences doubted something already well accepted by the academia, namely that the process of origin of the human species was similar to those of other species, under the action of evolutionary phenomena such as natural selection, mutation, genetic variation, founder effect, and others (Mirazón Lahr and Foley 1998).

It is important to point out that the Brazilian biology teachers' perception of the human origin is in line with that of most of the Brazilian population. A survey conducted by the Brazilian Institute of Geography and Statistics

(Instituto Brasileiro de Geografia e Estatística [IBGE]) revealed that 31 percent of Brazilians believe that God created human beings sometime in the last 10,000 years, despite existing scientific evidence. Some other data from the same survey may explain the unexpected percentages in the teachers' responses: over half the Brazilians (54 percent) believe that human beings originated through an evolutionary process guided by God as a greater force. We may conceive that there is no problem in this type of conception since it incorporates and accommodates both scientific and religious knowledge; however, it has potentially deleterious effects on secular and academic education. The first concern is the recent expansion of the Intelligent Design movement, which, in principle, would not be a problem, but is just one more pseudo-scientific approach spread especially through the Internet. However, we must point out that this movement both negates the principles put forward by the Darwinian evolutionary theory and asserts that various fields of science deceive students and the population as a whole. The second concern is the support by 89 percent of the Brazilian population of the teaching of creationism in all schools (Costa, Melo et al. 2011).

The general lines of the processes that led to the origin of the human species have been well elucidated, despite some gaps due to the lack of fossil records (Benton, Wills et al. 2000). The passage of hominids in Africa is relatively well documented (Conroy and Pontzer 2012). Therefore, we can argue that discussing whether the human origin was a natural process would be irrelevant and restricted to a few countries with a strong religious position to the point of affecting secular education, such as Brazil and the United States, which have seen strong creationist pressures, either in the form of the Young Earth creationism or disguised as Intelligent Design. However, the results of a study conducted in various countries that investigated whether human beings evolved from older animal species demonstrated that the origin of the human species continues to be a delicate subject to the population in general. Although in some countries, like Germany and France, this study found 74 and 80 percent agreement with the evolution of human beings from older animal species, many other countries had high indexes contrary to the natural origin of human beings, such as Greece (34 percent), Turkey (52 percent) and the United States (40 percent) (MacDowell 2011). These data show that it is important to investigate and discuss how biology teachers perceive the origin of the human species, since most of the population has contact with this subject only through teachers in secondary education.

The finding that nearly half of the Brazilian teachers interviewed in this study do not consider the human origin a natural process invites reflections on its possible impact on Brazilian education. First of all, it is necessary to consider that these teachers belong to and are influenced by a markedly religious society and that the influence of religions has greatly increased in the last years, including in political and educational organizations (Sepulveda 2004) and notably in official proposals of the Brazilian Council for the introduction of creationism in science classes in formal public education (Dorvillé and Teixeira 2015). Another aspect to discuss is the limited training of Brazilian biology teachers regarding natural sciences, the Darwinian theory and the origin of the human species, which results in a distorted conciliation of these themes and the teachers' religiosity (Rutledge and Warden 2000). It is important to evaluate whether this conception of special origin of the human species held by nearly half of the teachers interviewed reflects on their teaching practice, whether they take their vision of the special origin of the human species, as different from the other species and in contrast to scientific evidence, to the classroom. Its impact can also be evaluated when biological evolution is treated as a marginal subject, taught late in the school term or through assignments, or even totally neglected because of the uneasiness it causes to the teachers and the students (Oleques, Bartholomei-Santos et al. 2011).

This discussion on how biology teachers conceive human origin and why a significant number of teachers do not believe that the human species had a natural origin, as did all the other species, can be demonstrated by the analysis of the collective discourse of the semi-structured interviews carried out with the teachers. When the teachers were asked to talk about the difficulties of teaching biological evolution, they showed signs of how their perception was formed. One of the teachers pondered: "Then . . . if the teacher has questions, how can he or she explain it to the student?" This teacher was reluctant to acknowledge the human species as a product of the evolutionary process partially because he or she did not know the fundamentals for understanding the evolutionary theory well. This lack of knowledge, together with the teacher's inner conflict with his or her own religion, revealed in the interviews, makes working on this theme difficult, and ends up having a negative effect on the teaching of Darwinian evolution, as shown by another teacher's utterance: "Sincerely, I go over it very quickly." Hence, to avoid facing his or her own lack of knowledge and internal conflict, the teacher prefers to neglect biological evolution in the teaching plan of topics to be taught in class during the school year.

It is conceivable that a significant number of teachers who answered question A.33, and many of those interviewed, experienced a conflict between their religions and the natural origin of the human species. Nevertheless, the interviews suggest the possibility of Dialogue and Integration between religious and scientific knowledge in the imaginary on human origin and evolution, following Barbour's (1990) categories. This accommodation is not necessarily positive, because apparently the teachers need to deny some basic concepts of evolution, especially the role of chance in the evolution of the human species, possibly inserting some religious elements into the scientific theory to consolidate this pacification.

Is Human Origin the Objective of Evolution?

Question A.44, "The emergence of the human species (*Homo sapiens*) was the aim of the evolution of living species," had the following results: 5 percent totally agreed, 6 percent agreed more than disagreed, 10 percent disagreed more than agreed, 79 percent totally disagreed.

Question A.44 suggests that the appearance of the human species is the objective of evolution, indicating a teleological approach. This view of biological evolution is characterized by the use of words like *function*, *purpose*, and *objective* to explain evolutionary phenomena, as if they had taken place with some "specific end" (Mayr 1974). Many biologists, educators, and even supporters of the evolutionary theory have considered this approach reasonable as a linguistic didactic resource to explain, for example, how structures that led to flight appeared, as if there were an intrinsic objective in the evolutionary process (Ayala 1999).

On the other hand, teleological explanations are disregarded by many educators and by most of the evolutionary theory supporters. Therefore, most of those who study evolution and its teachings consider that this approach should be excluded from the school biology content. The teleological perception of evolution is nearly natural. It is easy to think as if there were a purpose or a clear direction; sometimes even scientists describe natural selection in this way, which contributes to disseminate this perception (Galli and Meinardi 2011). Teleological thought may seem instinctive; however, it is necessary to discuss whether biology teachers see evolution with this bias due to the strong influence of religion in countries like Brazil; especially in relation to human origin and evolution. The teachers' views will influence the students' concepts of this theme. When teachers have a teleological view, they may have difficulty in accepting the well-established scientific basic principles of the evolutionary theory, such as random mutation, chance, and genetic variation, among others. We can conclude that the teleological approach is one of the sources of the students' elaboration of equivocal concepts on biological evolution. Therefore, it is important to know how teachers deal with teleology in this respect (Tamir 1985).

Another aspect to consider is that the teleological vision may be used as an instrument in the diffusion of the intelligent design (ID) movement, which is reasonably popular and growing in Brazil (Silva 2017). Obviously, the assertion of a Creator as the conductor of the evolutionary process, as proposed by the ID, is not viable in a country that claims to have a secular education system. ID and any other type of creationism are not considered a scientific hypothesis and much less a theory to be taught along with the Darwinian theory of evolution or as an alternative to it (Scott and Matzke 2007). There currently exists an environment of change in the religious spectrum that may favor this type of approach, with an increasing number of Evangelicals in contrast to Catholics, which favor creationist and ID views and may affect politics and education (Silva and Mortimer 2014). If the teachers' understanding of evolution is influenced by creationism through ID, they may also adopt a teleological view, as evolution would have been driven by a Creator or Designer. For those who seek a conception that conciliates science and religion (Barbour 1990), the teleological approach may play this role, delivering a more palatable view of human origin and evolution.

None of these risks of the teleological interpretation of the human origin was found relevant in the sample of Brazilian teachers investigated. Nearly 90 percent of the biology teachers interviewed disagreed that the origin of the human being is the objective of evolution, which keeps away the risks of distortion and bias from a religious viewpoint. It is worth mentioning that 11 percent of the teachers interviewed agreed with the teleological viewpoint of human genesis, which might influence their teaching of human evolution. However, some aspects must be examined later: whether and how the teachers convey this conception to their classes and how this vision was constructed, whether out of limited scientific knowledge or by exposure to equivocal concepts in their training.

It was not clear from the teachers' discourse collected in the interviews whether the teachers understand the teleological perspective, but it was clear that some believe in a certain driving force, such as God, that afforded the appearance of human beings, which is indicative of this kind of approach. However, the greatest difficulty related to the teleological position was made evident in the teachers' account of student resistance to evolutionary principles, such as the role of chance. One of the teachers' discourses exemplified how the students tend to think of God as the creator of the human species, adopting this interpretation even when scientific arguments were presented in the classroom: "That the student brings a religious ideal of the creation of human beings in the image of God . . . And when you go into, into evolution, you have to show him/her another view, one that demonstrates where human beings came from and not what is in their own faith, from their own religious conviction . . . Then he is very attached to, to . . . puts his faith above anything else . . . And then, the human being ends up like, like being the creation of God anyway . . . You can't argue with that."

Considering Barbour's (1990) categories, if we take into account that the great majority of the teachers interviewed reported themselves as being religious, we may suppose that these teachers have an independent position when they analyze human origin from the teleological viewpoint. After all, if they believe in a supernatural force but do not consider that evolution was guided, an objective in evolution being the appearance of human beings, this means that they manage to have an opinion aligned with the Non-Overlapping Magisteria (NOMA) perspective, which proposes that religion and science are distinct domains and that their overlapping should be avoided whenever possible, respecting each one and refraining from conflict (Gould 1997a).

Is the Origin of Humans Associated with Religious or Scientific Expressions?

The percent results of question A.62, "Expressions that you think are the most strongly associated with the origins of humankind," were as follows: 5 percent Adam and Eve, 15 percent Australopithecus, 6 percent Creation, 29 percent Evolution, 18 percent God, and 27 percent Natural Selection.

The term "evolution" was the most relevant in the answers given by the teachers interviewed, which demonstrates that they perceive that this concept is fundamental for explaining human origin and that they agree with the pieces of evidence presented by the various fields of science, such as genetics, paleontology, and biostratigraphy (Cavalli-Sforza et al. 1988). It is important to point out that biology teachers need to let go of the idea that they believe in evolution, as was noticed in many interviews, because it is not a matter of belief. It is a theory based on proofs from various fields of science (Williams 2015).

The Brazilian teachers interviewed in this study cited that natural selection was one of the most important factors in the process that resulted in appearance of the human species, which agrees with Nehm and Reilly (2007). These authors stated that it is fundamental for teachers to understand the role of the phenomenon of natural selection so that they can explain and demonstrate its importance to their students in the classroom. This fundamental role of natural selection is supported by a set of pieces of scientific evidence ranging from fossil records to genetics (Stringer and Andrews 1988). Despite the relevance of natural selection, a significant number of Brazilian biology teachers interviewed did not relate human origin to human evolution, suggesting the existence of shortcomings in their training (de Freitas and Villani 2016), since the relevance of this evolutionary mechanism ought to be very clear for teachers in training. Besides being crucial to understanding the evolutionary process, natural selection is relatively simple and can be easily accepted by students (Via 2009).

Few teachers investigated indicated that Creation is relevant for the evolutionary process that resulted in the origin of humans; however, a significant number considered the role of God important. Jalil (2009) proposes that if the origin and evolution of humans is interpreted differently from those of other organisms, evolution in nature would be reasonably accepted, but would result in numerous obstacles and arrangements in relation to humans. Some prefer to think of God only in the beginning, while others think of God as having steered evolution toward His greatest creation, the human being, by producing mutations, infusing the spirit and working through miracles, for example. These teachers who mentioned Creation in the evolutionary process may be reasoning in this way to rationally accommodate their strong belief in a Creator and the knowledge of human evolution. Other studies have confirmed this need when they observed an impasse among university students to adapt their religious beliefs to the scientific view of human origin (Barnes, Keilholtz et al. 2008).

One fact that draws attention in the biology teachers' answers is their mentioning that the name of God is more significant to human evolution than the term Australopithecus. In what concerns scientific knowledge, there is no room for doubt that ancestral hominids from the Australopithecus genus make up part of the human phylogenetic tree (Wood 1992). However, the majority of the Brazilian teachers apparently did not agree with this perception. It seems complicated for Christians, as most of the teachers themselves reported, to accept basic fundaments that explain human origin, which leads to their resorting to a divine figure to explain it. Nevertheless, it has been demonstrated that when the Darwinian theory is presented with the exclusion of God, a sample of the Brazilian population rejected the evolutionary theory (Silva, Araújo et al. 2014). The discourse of one of the biology teachers illustrated clearly the problems they experience in their training in Brazil, that interfere with their understanding of the evolutionary and phylogenetic phenomena that resulted in the appearance of the *Homo sapiens*:

> Everything, they think, they think that Man came from monkeys; in fact, they, that is, there is a controversy; sometimes they say that Man came from Adam's rib and that the woman came from Adam's rib and that's all. Another point of view, another one, some also say that they believe that Man came from the monkey; then I ask them, if Man came from the monkey, why don't we have monkeys turning into humans these days.

This teacher confounds ancestor hominids with monkeys and deals with human evolution based on beliefs and has difficulty in articulating evolutionary theory in relation to human origin, demonstrating that attention should be paid to human origin during the training of science and biology teachers.

A relatively small number of teachers mentioned the biblical figures of Adam and Eve, even though these mythical characters appear in the Western Christian religious culture as fundamental in the origin of humanity (Livingstone 2011). The difficulty of these teachers to reconcile scientific knowledge with these mythical figures was evident in their interviews. One teacher mentioned what happened in a biology teacher training course: "They asked this, right, err—does Man come from monkeys? They do not make this connection yet between . . . Adam and Eve . . . and such, they do not make this connection, we realize that they get a bit confused." This teacher's utterance reveals a difficulty in reconciling a literal interpretation of the Bible, as well as limited knowledge and understanding of the evolutionary process that resulted in the human species on considering that humans came from monkeys. Another very experienced teacher revealed her doubts and confusion on articulating knowledge of human evolution and biblical characters. "Even so, when you deal with human evolution, if you take the Bible, then you get to Adam and Eve, Cain and Abel . . . so what? And the rest? Where did they come from?"

The analysis of the answers to question A.62 considering Barbour's (1990) categories revealed two ways that teachers perceive the relationship between science and religion and human origin. Many teachers mentioned terms like natural selection and evolution as fundamental for human origin in detriment to the mythical figures of Adam and Eve; therefore, signaling that they fall into the Independence category, as they indicate these alternatives, even though they considered themselves Christians. However, the Dialogue and Integration categories can also be considered for other answers, especially when various teachers perceived the role played by a divine figure in human origin. Finally, the Conflict category was also observed when the teachers hardly mentioned the Australopithecus genus as a member of the human phylogeny, accordingly, as being a part of the origin of the human being. We would have expected that a biology teacher who is able to harmonize his or her faith and religion with the scientific knowledge, would also consider the importance of the ancestral hominids.

Did the Human Species Result from Evolutionary Processes, Divine Work, or Both?

In relation to question B.28, "Which of the following four statements do you agree with the most?" The percent answers were: 44 percent It is certain that the origin of humankind results from evolutionary processes; 6 percent Human origin can be explained by evolutionary processes without considering the hypothesis that God created humankind; 29 percent Human origin can be explained by evolutionary processes that are governed by God; and, 21 percent It is certain that God created humankind.

The percentage of answers from the Brazilian teachers interviewed relating human origin to the religious field was high (50 percent). These data corroborate another study that reported that even among biological science students, future teachers of biology, the percentage of views of human evolution related to the religious field was high (Souza et al. 2009).

In the analysis of the teachers' answers on human origin, 21 percent of the teachers can be categorized as creationists, since they considered that God created humankind and, therefore, believe in a supernatural action in the creation of the human species (Engler 2007). These teachers did not consider biological evolution as playing a role in the process of human origin. We can thus conclude that they experience a conflict with their religious beliefs (Barbour 1990) in which they tend to place their faith above scientific knowledge when they express their view on this theme. It is difficult to know whether these teachers manage to reconcile their faith and science. However, their answer pattern leads us to suppose that they deny scientific concepts in relation to human origin. Therefore, we conclude that their perception falls into the category of conflict between science and religion (Barbour 1990).

A significant percentage of the teachers interviewed (29 percent) consider that although humans originated through evolution, it was conducted by God. These teachers may conceive human origin through so-called anthropocentric creationism, which accepts the theory of evolution, but claims that evolutionary processes advance according to God's command and that God creates Man's soul (Engler 2007). The teachers that gave this answer can be classified within the line of thought called Creationist Evolutionism, which accepts most of the principles of the theory of evolution, but also accepts the hand of the Creator as a participant (Scott 1999). We suppose that teachers who adhere to this type of evolutionary approach create a space for the expansion of the Intelligent Design movement in Brazil, because, for its followers, evolution was driven by God (the designer). Furthermore, they end up denying some of the main pillars of Darwin's evolutionary theory (Silva 2017). These teachers' answers can be classified as an attempt at dialogue and integration between scientific and religious knowledge (Barbour 1990).

The answer given by the teachers most frequently (44 percent) indicated that the human species originated through evolutionary processes. As the teachers generally claimed to be religious, we can suppose that they believe in an original force of creation, maybe of life, of Earth or the Universe, but that evolution followed its course driven by natural forces. This type of perception can be compared to that which is called Theistic Evolutionism. It seeks to reconcile with creationism without denying any of the principles of the Darwinian theory (Scott 1999). This line of thought suggests that God created the evolutionary process without interfering with the changes the beings involved underwent. These teachers can be classified being in the Independence category (Barbour 1990) because they manage to analyze this issue and separate science and religion into different magisteria.

Few of the teachers interviewed (6 percent) reported to believe that evolution gave origin to Man without God's interference, a view more aligned with current academic knowledge. Other studies demonstrate that, in Brazil, when the figure of God is removed from the evolutionary process, teachers tend to deny evolutionary theory (Silva, Araújo, et al. 2014). These teachers are probably ideating human origin within so-called Evolutionary Materialism or Atheistic Evolutionism, a non-religious perception (Scott 2009). In this view, science focuses on matter, energy, and their interactions. Furthermore, it is science's task to explain the natural world, not the supernatural one. In this view, science is neutral in relation to religion, neither against nor in favor of it, and supernatural phenomena are out of its scope, because materialist (scientificist or naturalist) philosophy dictates the rules of how to do science. The teachers who answered in this line of thought fit the Independence category (Barbour 1990), as they deny any interference from God in human origin and evolution, even though they are mainly religious.

The teachers' conflict in elaborating their concepts on human evolution is illustrated by one teacher's interview: "When you look at the stories, there were other animals before, before Adam and Eve . . . and other people . . . It's very, I think it is very complicated, like this . . . it is the most complicated part of science, it's evolution." This teacher expresses great difficulty in reconciling scientific knowledge and strong religious belief. Fundamental elements of evolutionary theory are also lacking, as the teacher shows limitations in the rationalization of the temporality of natural events. The result is that this teacher considers evolution the most difficult theme to understand.

Another teacher's statement demonstrates how it is often complicated to try to reconcile science and religion or to attempt a dialogue and integration: "Generally, some teachers say that they don't believe in God, they say that they believe only in evolution really. It simply was . . . No. No way . . . because in fact when they say it, I say, folks, who headed this evolution? It was something higher; what is this higher thing? For us, we say that it is God." This teacher's statement has two interesting elements, one related to the criticism of atheist evolutionary teachers by a teacher who works with evolution, and another that shows how the teacher's statement aligns with those of

Intelligent Design supporters. Thus, this teacher probably denies some of the basic suppositions of evolutionary theory, such as random mutations, chance, genetic variation, and others.

Does the Darwinian Evolutionary Theory Contradict Your Beliefs?

Question B.29a, "The theory of evolution contradicts my own beliefs," had 23 percent agreement and 77 percent disagreement among the teachers interviewed.

Despite being religious, most teachers (77 percent) fall in the Independence category and do not consider the evolutionary theory to be opposite to their beliefs, nearly a quarter of the teachers interviewed (23 percent) indicated that evolutionary theory contradicts their personal beliefs. These teachers experienced what Barbour (1990) calls the conflict between science and religion. Even though they are biology teachers and evolutionary theory is the guiding thread in this course, a fundamental one for its understanding (Dobzhansky 1973), this topic seems to be opposite to their faith. This percentage corroborates the results reported by Oleques and colleagues, who also found that many biology teachers refute biological evolution as a theme, both conceptually and in their teaching practice (Oleques, Bartholomei-Santos et al. 2011).

The interviews revealed an obstacle in dealing with teaching evolution and the questions that it raises in relation to the religions of the teachers themselves. The teachers' unease was clear in dealing with evolution, especially human evolution and knowledge from the Christian Bible. One teacher remarked on the experience with colleagues in the biology undergraduate course: "However, in the undergraduate course I had religious colleagues most . . . most often Protestants, who opposed the teachers' statements using Bible data." This type of discourse agrees with the findings from an investigation by Sepúlveda and El-Hani (2004) of the training of biology teachers of Protestant faith. One of the ways these researchers observed that Evangelical teachers use to deal with biological evolution is the absolute and systematic refusal of the theme, demonstrating an aversion to it. This investigation reveals that a certain misunderstanding of the Darwinian theory by biology teachers may be the result of major conflicts that they experience between the scientific explanation of evolutionary processes and their religious conceptions (Rosa et al. 2002).

We need to reflect on the impact of the possible negation of evolutionary theory by biology teachers on the Brazilian population, especially students. Most students will have contact with evolutionary theory only in secondary education, where the investigated teachers work. Therefore, if the teachers deny this theme, the students are also denied contact with a phenomenon so fundamental to various fields of science, such as agriculture, medicine, and others (Purushotham and Sullivan 2010).

Does Creationism Contradict Your Beliefs?

In question B.29b, "Creationism (including the creation of human beings by God) contradicts my own beliefs", 34 percent of the teachers interviewed agreed and 66 percent disagreed.

Two out of every three Brazilian biology teachers in our sample considered that Creationism does not contradict their beliefs. If we relate the prevalent answer to this question and the answer of the previous question one can perceive a contradiction, because apparently it would not be possible to believe in evolution and be a creationist. However many authors do not perceive this contradiction because it would be possible to believe in a creator and at the same time to recognize that the evolutionary process occurred with all its presuppositions. It is even useful for a religious person to have evolutionary knowledge, because the incoherence of the existence of imperfections and sufferings in nature before an almighty God, would be explained through natural selection (Ayala 2007).

The results of the present study were opposite to those of investigations carried out with Argentinean and Uruguayan teachers (Silva 2015). Only in the study carried out in Brazil did most teachers support the creationist viewpoint. This result invites reflection on how Brazilian biology teachers deal with Evolution as a subject in class, such an important theme for the understanding of biology. Further investigation is needed to measure its impact. Apparently, the teachers experience a conflict regarding their scientific knowledge and later move toward dialogue and interaction between science and religion (Barbour 1990).

The collective subject technique (Lefevre and Cavalcanti Lefevre 2014), through first-person discourse, used to demonstrate the synthesis of internal conflict that teachers experience, shows that there is a complex combination of scientific and religious knowledge: "I experience a very complex internal conflict; on the one hand, I know that I must teach evolution, but on the other hand I have questions on this subject. What are my students going to say? When I try to reconcile the scientific knowledge of human origin, for example, in relation to primitive primates, I struggle with my religious knowledge learned from the Bible. I find it difficult to discuss polemical issues of evolution with the students because I don't have arguments; I have not learned them in my training. I try to bring the divine figure into the subject of evolution to reconcile the two fields of knowledge."

The teachers' discourse reveals a series of questions in relation to biological evolution. Religion is a major factor in the raising of these questions and the acceptance or rejection of this topic (Trani 2004). A fundamental issue is whether these questions raised based on religion, and the effect of the teachers' limited training on evolutionary theory in their teaching practice, contribute to the formulation of a science-religion hybrid position that impoverishes the classes on this subject.

The fact that biology teachers are supportive of creationism, together with an increase in the political and educational influence of Evangelicals within the current Brazilian religious spectrum, may have contributed to a number of attempts to introduce religious education in Brazil (Silva and Mortimer 2014). This effort is realized in the proposals of mandatory teaching of creationism presented to the Brazilian Congress. The number of attempts to introduce creationism in education has been great, especially in the state of Rio de Janeiro (Sepulveda 2004). In Brazil, the Intelligent Design view has been divulged by scientists at universities and by the media in general (Silva 2017). Even the federal administration itself, through the Ministry of Human Rights, a dyed-in-the-wool Evangelical herself, publicly expressed that science should not be left on the hands of scientist and that was why the theory of evolution got into the schools (https://www.youtube.com/watch?v=ImgJ5c6-ock). The minister's statement indicates that Evangelicals and creationism may be a barrier to scientists and the teaching of evolution, which is a concern of most of the teachers interviewed.

Has God Participated in the Evolution of Species?

When presented with question B.48, "Indicate your evaluation of the importance of the God in species evolution," the teachers answers were as follows: 50 percent Great importance; 21 percent Some importance; 14 percent Little importance; and 15 percent No importance at all.

Most of the Brazilian teachers interviewed (71 percent) considered the participation of God in the evolution of the species to be important. As demonstrated by the previous question, the teachers support the creationist viewpoint, for which the role attributed to God in the evolutionary process is to be expected. We can consider the hypothesis that these teachers carry their religious beliefs over into the approach of biological evolution, which may also interfere with their learning of the subject (Souza 2007). Therefore, it is possible that if the teachers accept some supernatural participation in evolution, their position my affect how students learn it.

Nevertheless, it is difficult to evaluate how the teachers' positions affect the students. The students probably receive a set of influences from their families, their religion, and from society as a whole. However, there is no doubt that teachers who combine their religious beliefs with the course content may affect the way the students conceptualize evolution. The interview of one of the teachers reveals how the students' perception of the origin of life and Darwinian evolution is influenced by this set of factors. "The students are not going to believe this because they have learned that creationism is part of the origin of life and all, since they were kids." Concepts religious in nature lead students to interpret biblical texts literally, taking them closer to an approach more related to creationism of the Young Earth (Engler 2007; Scott 1999), which suggests that life and Man were created recently, as stated in the Holy Scriptures. One of the interviewees reported how the students negated evolution based on this literal interpretation of the Bible: "That's it, they think that God created [Man] such as is in the Bible, literally; they don't believe natural evolution happened."

The synthesis of the collective discourse (Lefevre and Lefevre 2014) of the teachers interviewed shows that they perceive this construction of an obstacle to the learning of biological evolution:

> What I consider the greatest obstacle to the teaching of biological evolution is that it clashes with the students' religiosity. Creationism, in its most radical form, as a literal interpretation of the Bible, also constitutes an aversion to the scientific view of the origin of Man and our evolution, another great obstacle that comes up in class. One of the aspects I notice that makes it more difficult is the increasing number of Evangelical students who have trouble reconciling the two types of knowledge, scientific, and religious.

A comparative investigation of the conceptions of Argentinean and Uruguayan teachers (Silva 2015) showed that teachers from these countries have a completely antagonist perception regarding the participation of God in the evolutionary process; almost no teacher considered a divine action. These teachers fall into Barbour's Independence category, while the Brazilian teachers swing between the conflict and dialogue categories and eventually reach a supposed integration (Barbour 1990).

FINAL CONCLUSIONS

Our results generally show teachers' difficulties in dealing with and limited of knowledge about biological evolution, as well as an attempt to compensate for this limited knowledge based on their beliefs. This probably allows the introduction of creationism and intelligent design in their teaching. This is a flawed attempt to accommodate these two fields, because it produces a confusion between science and religion (Collins 2003). The interviews and analysis of the collective discourse of the teachers' statements revealed two obstacles to teaching evolution in Brazilian schools. One of them is the internal conflict experienced by the teachers in relation to biological evolution, because it conflicts with their beliefs and there is no way to solve it easily, especially due to the limitations of scientific knowledge. The other obstacle revealed is the conflict with the students' religiosity, since they would be refractory to Darwinian evolution, especially in relation to human origin and evolution, and teachers would not be able to meet this demand. The analysis of the set of responses of the teachers to the questionnaire and in the interviews revealed what the teachers' main conceptions of biological evolution are. The teachers have shown that it is difficult to separate scientific and religious views, so they do not demonstrate independence between the fields. Initially they experience conflict between religion and science, seeking a dialogue and finally trying to integrate the different perspectives. The types of religion professed by the teachers proved to be important in the elaboration of this course of conceptualization.

We should note in the data an apparent contradiction, since the same teachers who believe in God, and consequently in creationism, also point out answers related to the scientific argument, in agreement with evolutionary theory. Nevertheless, for many prominent evolutionists, the conflict between creationism and evolutionism is unnecessary and unproductive, for the theory of evolution does not directly address the origin of life; it is about the modifications of species over time. Creationism, on the other hand, focuses on the emergence of life. The two conceptions deal with distinct things, so it stands to reason that the teacher can be creationist but at the same time could conceive evolution and teach it (Gould 1997b).

To further highlight this contention that there may be no contradiction in the seemingly disparate views, it is interesting to mention that Darwin himself said in one of his letters, in a forceful manner, that "It seems to me absurd to doubt that a man may be an ardent Theist & an evolutionist."[1] In addition, one of the greatest current scientists in the field of biology and leader of the Human Genome Project, Francis Collins, is an assumed creationist, but at the same time a strong supporter of the evolutionary theory and critical of intelligent design theory, because the supporters of this hypothesis deny evolutionary assumptions (Collins 2003). Finally, an influential organization of Evangelicals in the United States, called BioLogos, constituted of convinced creationists, recognize the value and truthfulness of evolutionary theory, developing ways to accommodate the two apparently antagonistic visions through the so-called theistic evolutionism (Luskin 2014).

Further studies are needed to detect the extent to which the findings of this study reach evolution classes in daily life, but it is possible to draw some conclusions. The perceived Brazilian context leads to difficulty in understanding and learning Darwinian evolution, which is fundamental to the understanding of biology as a whole. Therefore, it is imperative to improve teachers' knowledge on the topic through professional development programs in a continuous way. Finally, it is also important to point out that when teachers use their personal beliefs to compensate for knowledge limitations, the religious diversity of Brazilian students tends to be disrespected, or at least biased toward a particular religious perspective.

NOTE

1. Letter no. 12041, available at http://www.darwinproject.ac.uk/DCP-LETT-12041.

REFERENCES

Abrantes, Paulo, and Fábio Portela Lopes de Almeida. 2006. "Criacionismo e Darwinismo Confrontam-Se Nos Tribunais . . . Da Razão e Do Direito." *Episteme* 11 (24): 357–402.

Ayala, Francisco J. 1999. "Adaptation and Novelty: Teleological Explanations in Evolutionary Biology." *History and Philosophy of the Life Sciences*. JSTOR, 3–33.

Ayala, Francisco J, and John C Avise. 2009. "Darwin at 200." In *Proceedings of the National Academy of Science*, 106: 2475–76.

Ayala, Francisco José. 2007. *Darwin's Gift to Science and Religion*. Washington, DC: Joseph Henry Press. https://doi.org/10.17226/11732.

Barbour, Ian G. 1990. *Religion in an Age of Science*. San Francisco: Harper & Row.

Barnes, Ralph M., Lesleh E. Keilholtz, and Audrey L. Alberstadt. 2008. "Creationism and Evolution Beliefs among College Students." *Skeptic (Altadena, CA)* 14 (3). Skeptics Society and Skeptic Magazine: 13–17.

Behe, Michael J. 1996. *Darwin's Black Box: The Biochemical Challenge to Evolution*. New York: Free Press.

Benton, M. J., M. A. Wills, and R. Hitchin. 2000. "Quality of the Fossil Record through Time." *Nature* 403 (February). Macmillan Magazines Ltd.: 534. https://doi.org/10.1038/35000558.

Carvalho, Graça Simões, and Pierre Clément. 2007. "Projecto 'Educação Em Biologia, Educação Para a Saúde e Educação Ambiental Para Uma Melhor Cidadania': Análise de Manuais Escolares e Concepções de Professores de 19 Países (Europeus, Africanos e Do Próximo Oriente)." *Revista Brasileira de Pesquisa Em Educação Em Ciências* 7 (2).

Cavalli-Sforza, Luigi Luca, Alberto Piazza, Paolo Menozzi, and Joanna Mountain. 1988. "Reconstruction of Human Evolution: Bringing Together Genetic, Archaeological, and Linguistic Data." *Proceedings of the National Academy of Sciences* 85 (16). National Acad Sciences: 6002–6.

Collins, Francis S. 2003. "Faith and the Human Genome." *Perspectives on Science and Christian Faith* 55 (3): 142–53. http://www.asa3.org/ASA/PSCF/2003/PSCF9-03Collins.pdf.

Conroy, Glenn C, and Herman Pontzer. 2012. *Reconstructing Human Origins: A Modern Synthesis*. New York: W. W. Norton & Company.

Cornish-Bowden, Athel, and Maria Luz Cárdenas. 2007. "The Threat from Creationism to the Rational Teaching of Biology." *Biological Research* 40 (2). Sociedad de Biología de Chile: 113–22.

Costa, L. O., P. L. C. Melo, and F. M. Teixeira. 2011. "Reflexões Acerca Das Diferentes Visões de Alunos Do Ensino Médio Sobre a Origem Da Diversidade Biológica." *Ciência & Educação* 17 (1): 115–28.

Dobzhansky, Theodosius. 1973. "Nothing in Biology Makes Sense Except in the Light of Evolution." *The American Biology Teacher* 35 (125–29). https://doi.org/10.2307/4444260.

Dorvillé, Luís Fernando M., and Pedro Teixeira. 2015. "O Crescimento Do Criacionismo No Brasil: Principais Influências e Avanços Recentes." *X Encontro Nacional de Pesquisa Em Educação Em Ciências*, 24–27.

Engler, Steven. 2007. "Tipos de Criacionismos Cristãos." *Revista de Estudos Da Religião* 7: 83–107.

Freitas, Denise de, and Alberto Villani. 2016. "Formação de Professores de Ciências: Um Desafio Sem Limites." *Investigações Em Ensino de Ciências* 7 (3): 215–30.

Galli, Leonardo Martín González, and Elsa N. Meinardi. 2011. "The Role of Teleological Thinking in Learning the Darwinian Model of Evolution." *Evolution: Education and Outreach* 4 (1). Springer: 145–52.

Gould, Stephen Jay. 1997a. "Nonoverlapping Magisteria." *Natural History* 106 (2): 16–22.

———. 1997b. "Os Três Aspectos Da Evolução." In *As Coisas São Assim—Pequeno Repertório Científico Do Mundo Que Nos Cerca.*, edited by Suzana Sturlini Couto, John Brockman, Katinka Matson, 95–100. São Paulo/SP: Companhia das Letras.

Graebsch, Almut. 2006. "Polish Scientists Fight Creationism." *Nature* 443 (October). Nature Publishing Group: 890. http://dx.doi.org/10.1038/443890c.

Jalil, Ana María. 2009. "El Debate Creacionismo-Evolución En Profesores de Biología y Al Interior de Las Clases de Una Escuela Confesional." *Revista de Educación En Biología* 12 (2): 61.

Lefevre, Fernando, and Ana Maria Cavalcanti Lefevre. 2014. "Discurso Do Sujeito Coletivo: Representações Sociais e Intervenções Comunicativas." *Texto & Contexto Enfermagem* 23 (2). Universidade Federal de Santa Catarina.

Livingstone, David N. 2011. *Adam's Ancestors: Race, Religion, and the Politics of Human Origins*. Baltimore: Johns Hopkins University Press.

Luskin, Cascy. 2014. "The New Theistic Evolutionists: BioLogos and the Rush to Embrace the 'Consensus.'" *Christian Research Journal* 37 (3): 32–41.

MacDowell, João A. 2011. "Evolução VERSUS Criação: Falso DilEma." *Veritas (Porto Alegre)* 56 (2).
Malafaia, Silas. 2009. "Criação x Evolução: Quem Está Com a Verdade." *Rio de Janeiro: Editora Central Gospel.*
Mariano, Ricardo. 2013. "Mudanças No Campo Religioso Brasileiro No Censo 2010." *Debates Do NER* 2 (24): 119–37.
Mayr, Ernst. 1974. "Teleological and Teleonomic, a New Analysis." In *Methodological and Historical Essays in the Natural and Social Sciences*, 91–117. Boston: Springer.
Mirazón Lahr, Marta, and Robert A Foley. 1998. "Towards a Theory of Modern Human Origins: Geography, Demography, and Diversity in Recent Human Evolution." *American Journal of Physical Anthropology: The Official Publication of the American Association of Physical Anthropologists* 107 (S27). Wiley Online Library: 137–76.
Nehm, Ross H., and Leah Reilly. 2007. "Biology Majors' Knowledge and Misconceptions of Natural Selection." *AIBS Bulletin* 57 (3). American Institute of Biological Sciences: 263–72.
Oleques, Luciane Carvalho, Marlise Ladvocat Bartholomei-Santos, and Noemi Boer. 2011. "Evolução Biológica: Percepções de Professores de Biologia." *Revista Electrónica de Enseñanza de Las Ciencias* 10 (2): 243–63.
Purushotham, A. D., and R. Sullivan. 2010. "Darwin, Medicine and Cancer." *Annals of Oncology* 21 (2): 199–203. https://doi.org/10.1093/annonc/mdp537.
Rosa, V., Elza C. N. Muniz, A. P. C. Carneiro, and Lidiane Goedert. 2002. "O Tema Evolução Entre Professores de Biologia Não Licenciados–Dificuldades e Perspectivas." *Encontro Perspectivas do Ensino de Biologia* 8.
Rutledge, Michael L., and Melissa A Warden. 2000. "Evolutionary Theory, the Nature of Science & High School Biology Teachers: Critical Relationships." *The American Biology Teacher* 62 (1). BioOne: 23–31.
Scott, Eugenie C. 1999. "The Creation/Evolution Continuum." *Reports of the National Center for Science Education* 19 (4): 16–17.
———. 2009. *Evolution vs. Creationism: An Introduction.* Vol. 62. Berkeley: University of California Press.
Scott, Eugenie C., and Nicholas J. Matzke. 2007. "Biological Design in Science Classrooms." *Proceedings of the National Academy of Sciences of the United States of America* 104 Suppl: 8669–76. https://doi.org/10.1073/pnas.0701505104.
Sepulveda, Claudia. 2004. "Movimento Criacionista: Um Risco à Formação Científica e Cultural Dos Alunos Da Rede Pública Carioca." *Boletim Da SBEnBio* 1: 4–9.
Sepulveda, Claudia, and Charbel Niño El-Hani. 2004. "Quando Visões de Mundo Se Encontram: Religião e Ciência Na Trajetória de Formação de Alunos Protestantes de Uma Licenciatura Em Ciências Biológicas." *Investigações Em Ensino de Ciências* 9 (2): 137–75.
Silva, Heslley Machado. 2015. "Professores de Biologia e Ensino de Evolução: Uma Perspectiva Comparativa Em Países Com Contraste de Relação Entre Estado e Igreja Na América Latina." Tese apresentada a Faculdade de Educação da Universidade Federal de Minas Gerais. Belo Horizonte.

Silva, Heslley Machado, Elaine Nicolini Nabuco, de Araújo, and Paloma Rodrigues Silva. 2014. "Perception of Brazilian Citizens about Darwinism and the (Non) Influence of God in the Evolutionary Process." Istanbul: Proceedings of INT-CESS14- International Conference on Education and Social Sciences P. https://www.researchgate.net/publication/275100477.

———. 2017. "Intelligent Design Endangers Education." *Science* 357 (6354): 880.1-880. https://doi.org/10.1126/science.aao3245.

Silva, Heslley Machado, Graça S. Carvalho, Paloma Rodrigues Da Silva, Daiana Evilin Gibram, Estadual Paulista, and São Paulo. 2016. "A Percepção Sobre a Hipótese Do Design Inteligente No Brasil A Percepção Sobre a Hipótese Do Design Inteligente No Brasil (Minas Gerais)." *Revista Conexão Ciência I* 11.

Silva, Heslley Machado, and E. F. Mortimer. 2014. "Rescuing Darwin Brazil." In *Latin American Perspectives on Science and Religion. International Perspectives on Science, Culture and Society*, edited by Ignacio Silva, 1st ed., 1:208. London: Routledge.

Souza, Carina M. A. 2007. "Presença Das Teorias Evolucionistas e Criacionistas Em Disciplinas Do Ensino Médio (Biologia, Geografi Ae História)." *Simpósio de Pesquisa Em Ensino e História de Ciências Da Terra. Campinas: UNICAMP*, 41–44.

Souza, R. F. de, M. de Carvalho, Tiemi Matsuo, and Dimas A. M. Zaia. 2009. "Evolucionismo x Criacionismo: Aceitação e Rejeição No Século 21." *Ciência Hoje* 43 (256): 36–45.

Stringer, Chris B., and Peter Andrews. 1988. "Genetic and Fossil Evidence for the Origin of Modern Humans." *Science* 239 (4845). American Association for the Advancement of Science: 1263–68.

Tamir, Pinchas. 1985. "Causality and Teleology in High School Biology." *Research in Science and Technological Education* 3 (1). Taylor & Francis: 19–28.

Tidon, Rosana, and Eli Vieira. 2009. "O Ensino Da Evolução Biológica: Um Desafio Para o Século XXI." *ComCiência*, no. 107. SciELO Brasil: 0.

Trani, Randy. 2004. "I Won't Teach Evolution; It's against My Religion. And Now for the Rest of the Story. . . ." *The American Biology Teacher* 66 (6). BioOne: 419–27.

Valença, Cristiana Rosa, and Eliane Brígida Morais Falcão. 2012. "Teoria Da Evolução: Representações de Professores Pesquisadores de Biologia e Suas Relações Com o Ensino Médio." *Revista Electrónica de Enseñanza de Las Ciencias* 11 (2): 471–86.

Via, Sara. 2009. "Natural Selection in Action during Speciation." *Proceedings of the National Academy of Sciences* 106 (Supplement 1). National Acad Sciences: 9939–46.

Williams, James D. 2015. "Evolution versus Creationism: A Matter of Acceptance versus Belief." *Journal of Biological Education* 49 (3). Taylor & Francis: 322–33.

Wood, Bernard. 1992. "Origin and Evolution of the Genus Homo." *Nature* 355 (6363). Nature Publishing Group: 783.

Chapter Four

Brazilian High School Teachers' Approaches to and Departures from Scientific Knowledge When Teaching Evolutionary Theories

Rita Tatiana Cardoso Erbs
and Olma Karoline Cruz de Medeiros

The initial motivation of our work originated from Alandeom W. Oliveira and Kristin L. Cook's study on the rise of creationism in the teaching of biology in Brazil. It helped us realize the possibility of allying the experience of a biology teacher with the experience of an educational psychology professor. Each one of us works with concepts of the evolutionary theory in our teaching environment: one teaching adolescents in regular high school and the other training teachers at a university. In order to understand the rise of the creationist theory in the teaching of biology in Brazil, we started collecting useful materials in the form of textbooks, theses, dissertations, and articles that revealed how evolutionary theories are approached and understanding how such scientifically accepted knowledge is still met with so much resistance when being taught and when being learned.

Evolution is the study of the transformations that occur across generations in physical as well as ethological transformations. Evolution does not always mean the victorious advent of beneficial traits in the species. It is a fact that these changes occur in all species. However, "how they occur" has been the subject of scientific inquiry for a long period of time. The religion vs. science conflict has always hovered over this work, and even today it causes conflicts between students and teachers.

Four main theories are typically presented in high school biology textbooks of students to explain the evolution of the species. They are: Creationism, Lamarckism, Darwinism, and Neo-Darwinism. Further, creationism is present in the textbooks of renowned textbook authors such as César and Sezar (2013), Amabis (2014) and Lopes (2015). These authors produce their didactic material recommended by the Brazilian Ministry of Education as part of the book collection of the National Textbook Plan (i.e., Plano Nacional do Livro Didático).

We noticed that the textbooks separate the theories about life into creationism and evolutionism. This division already constitutes an oversight since creationism is presented as a scientific theory rather than an explanation of legendary or mythological nature about the inception of life. The epistemological effect is that creationism is rendered equal epistemological status as an accepted theory in the scientific realm, for it is presented alongside theories that are part of the scientific knowledge accumulated over the years of investigation. The strong influence of faith and religious belief pervades not only in biology textbooks published in Brazil but also much of the 159 years of scientific discussion and refinement that followed Darwin's proofs presented in *The Origin of Species*, which was published in 1859.

Although creationism is a rejected explanation in academia, for it asserts that a "higher force" created all species as they are, this explanation is defended, among others, by Western Christian Jewish religions, which do not believe in the transformation of species through generations. That is, some defend Fixism, the theory that species do not change and are fixed; they do not evolve. Today, there is countless evidence of species' transformism, a fact evidenced by the study of fossils, comparative anatomy, embryology, homology, analogy, and biogeography.

Evolutionary theories that are presented in high school textbooks start with Lamarckism, which is approached in textbooks with a scientific focus, with the understanding that species change according to the laws of use and disuse and the transmission of acquired characters. Even though this theory was already surpassed scientifically, it was the object of studies and investigation for the construction of later theories. "Although other scientists have already raised ideas about the processes of change in living organisms over time, challenging the immutability of species, it was from Darwin's work that we began to accept the fact that species change over time, originating other species" (Lopes 2015, 334). While Darwinism is presented as a theory that defends evolution by natural selection, contesting any legendary or mythological explanation, neo-Darwinism is presented as the junction between Darwinism and new genetic discoveries such as gene recombination and mutations in DNA.

Given the above, some questions arise about how this content is being applied in schools and how students are building their knowledge on this subject. We also inquire about how teachers approach evolutionary theories and the creationist explanation in the classroom, as well as their training and positioning when working with this content. From the reflections presented below, we identify how much this study is in an intricate field, since it is evident there is a gap between scientific acceptance and religious explanations. How do we then comprehend the rise of creationism, given the wealth of evidence to support evolution as a scientific theory? Could this view be addressed only in a historical context in which man sought explanations for natural phenomena in legends, myths, and gods? Other beliefs are also based on these same precepts to explain the origin of man, universe, and Earth.

Perhaps in Brazil, Christian Judaic ideologies have found a great receptivity of the population to perpetuate contents and ideas that have already been scientifically clarified. The teaching of evolution, due to the ideological strength of creationism in the population, can cause the teacher to develop a confrontational and defensive discourse trying to address students' misconceptions. Often, the training of the teacher does not aid or strengthen the teacher so that he can perpetuate the scientific issues even if they are contrary to the position of the population in general.

Formative reality assumes that by having scientific knowledge about content, the teacher will be able to assist the student in the construction of hypothetical deductive thinking, as if resistances and barriers did not exist, but, in fact, they do exist as presented by Cicillini (1997) in his thesis. Cicillini analyzes the forms of representations that teachers have on the content of evolution, presenting fragilities in relation to ideological connotations, fragmentation and distortion of content, dubious language, simplification of concepts and misused verb tenses in speech. In this same perspective, the author states that:

> We also observe that the words associated with the evolutionary process like *adaptation* and *evolution* are used with different connotations, contrary to those used by science. Some factors that may lead to such conceptions are: the polysemy of words, the simplistic notion of natural processes, and an anthropocentric view of evolutionary processes. These factors deviate teachers under the conceptual point of view, from the view shared by the current scientific community. In this sense, the teaching-learning process of biological evolution is impaired, since teachers do not adequately master some evolutionary concepts. (Oleques 2011, 260)

METHODOLOGY

Given the relevance of this theme, we used interviews with high school teachers to reveal how evolution content is approached as well as the teachers' areas of emphases and their vulnerabilities. To do so, an (auto)biographical research approach was taken with 10 teachers of public schools. The production of the data was carried out through semi-structured interviews on the theoretical basis of Bourdieu's (1989) and Belei's (2008) work, who emphasize the complexity and delicacy of approaching the research issues based on the life history of the teacher. Because of the rapprochement between the researcher and the participant and the theme involved, possible divergent perceptions emerge that can connote, like said by Bourdieu (1989), as symbolic aggressions. From their responses, we wrote a report that describes which dimensions appear in the teachers' narratives and explanations about the evolutionary theories. The interviews were conducted in three modules: identification, teaching methodology, and development of the theme in the life history of the teacher.

Regarding the identification data, information was collected on the academic formation of the interviewee: time of action, time in which the teacher has been working and experience with the subject. Subsequently, questions were raised about the teaching methods in the approach to the subject, such as the use of images, slides, texts, investigative works, university exam exercises and reading. Finally, data were collected on how the teachers prepared themselves about the construction of the ideas related to the subject, listing their development and the development of the evolutionary concepts in their life history, considering religiosity, family, and the receptivity of the scientific approach.

For the present study, six biology teachers—five women and one man—were interviewed. The female majority was due to the primacy of women in the practice of teaching in Catalão. All the teachers interviewed were from the public network, all are older than thirty years and have been teaching for more than twelve years. All had children and declared themselves married. The receptivity of the interview was positive, possibly because the interviewer was also a biology teacher and active in the state and private network of the city of Catalão-Goiás. Further, pseudonyms were used to protect their personal identities.

The scheduling of the interviews counted on the flexibility of the interviewer's schedules and locations, since the availability of the teachers' time for the interviews was limited. All the teachers work forty hours a week or more and still must dedicate themselves to their home and their families. Thus, the interviews were carried out in the schools of the teachers and some later communication was made by telephone with the recording of audio. All of the teachers signed an informed consent form, where they reported being older than eighteen and accepted their participation in the voluntary study. They were duly informed and clarified by the responsible researcher (Olma Karoline Cruz de Medeiros) about the research procedures and methods involved as well as the possible risks and benefits of participating in the study. It was guaranteed to everyone that they could withdraw their consent at any time, without leading to any penalty.

One procedure that had to be used in this research was the protection of interviewees' names. All names were changed, and for better identification we chose simple names that identified only the gender of each of the interviewees. With the interviews, we opted for the method of content analysis for the treatment of the data and information contained in the material, using as a methodological reference (Bardin 2016). Content analysis is how to process information, attitudes, or topics contained in a message or in a document. As Erbs and Abrahão (2018) point out, Bardin defines content analysis as a set of communication analysis techniques, aiming to obtain, through systematic and objective description procedures of the messages, indicators, both quantitative and qualitative, that allow inferences regarding the conditions of production and reception of such messages.

As methodological steps, the author describes the following phases: pre-analysis; material analysis; treatment of results, inference and interpretation. In the first phase, the organization of the material was done to elaborate a schematic, although flexible, precise of operationalization and systematization of the ideas, effecting: superficial reading of the material (with the goal of, when having an initial contact with the material of analysis, the structure of the text, to recognize the most used concepts, to know, preliminarily, positions about the important research questions and the context of the interviews).

In the second phase, the material was analyzed, which consisted of the codification and categorization of the material. Codification here means the systematic transformation and grouping of the information into units as to enable a description of the relevant characteristics of such data. In the interpretation phase, the content that emerged in the previous phases will be reviewed, interweaving theoretical and practical aspects according to the subjectivity of the authors of the present investigation and the purpose for which the research is intended.

We read and reread the testimonies to perceive recurring subjects, dilemmas, social contexts, metaphors used, and all resources that orality makes possible to the author. From this interpretation, we wrote the notes that gave rise to the summaries of each of the statements, the choice of the depositions of the deponents and the dimensions presented by the authors that will enable the construction of a meta-narrative in relation to the theme of the present research.

KEY FINDINGS

Teacher Alice: "God and Science are almost the same thing."

Alice is thirty-eight, married with two children—an eighteen-year-old daughter and an eight-year-old son. She has professional experience as a teacher for nineteen years. When questioned about her religion, she spoke of being catholic since she was born. She descends from a fervent Catholic family, nevertheless her father today is Protestant. She has been a music teacher since she was fifteen. She plays the saxophone; her father was a conductor.

On the teaching of biology, she reported that the subject she likes the most is genetics. She is very fond of mathematics, and currently she is pursuing another degree in environmental engineering. She started her civil engineering degree but dropped out after three semesters. In her teaching practice, she had already taught chemistry and mathematics classes. Also, when referring to the subject that she has less affinity to, she mentioned botany, although she really likes the environment.

When referring to evolutionary theories, Alice mentioned that she likes to work the evolution subject addressing the issue of prejudice and races in the human species. According to the teacher, we are all evolutionarily equal, emphasizing the importance of demystifying issues of race superiority in the classroom. As a clarification, the word race is used in the teaching of biology as a function of specific terms that are studied in the contents of taxonomy and classification of species, this term is used to cite subspecies as in the case of dogs, cats and cattle.

Teacher Alice has a concern with scientific production in the area, she wrote an article: "Reflecting Racism in Biology Classes" and according to it, science teaching should include understanding the social function and should address racial issues. Her classes on evolutionary theories begin with the following sentence: "Is the grasshopper green because it lives in the grass or it lives in the grass because it is green?" Using this to introduce the theories of Lamarckism and Darwinism, because the question refers to the two theories, the students have the opportunity to begin their reflections. She also likes telling theories as if they were a story, claiming it to be a process that is happening every day. She uses the example of the moths of Manchester, being a well-known example of natural selection that occurred before and after the industrial revolution, always present in the didactic books.

In relation to God, Alice said that she approaches science by understanding that they are the same thing, saying that there is no confrontation between the teachings of the Catholic Church with scientific evidences, and affirms that God and Science are the same with different explanations. God could have created natural selection and evolution. She states that science shows the same context of God only in a more scientific way. Nevertheless, she says that she is not much questioned about God by the students.

Talking about the evolutionary theory, she states that a theory presupposes hypotheses and evidences. She reads excerpts from Darwin's book, *The Origin of Species*, and excerpts from other scientists of the time taken from an ancient didactic book by author Sergio Linhares, which contains interesting fragments of this content. She places creationism, Lamarckism, and Darwinism all as theories. She also reports that her mother was a teacher and had always learned that scientific knowledge is very important and again reinforces that they—scientific and religious knowledge—are the same thing.

Teacher Liliane: "Being a scientist does not rule out believing in God."

Liliane is forty-three years old, married, has a child, and has been teaching for fifteen years. When she was asked about religion, she identified herself as a Spiritist. About what she likes to teach, she referred to genetics, arguing that students are very interested in this, which makes the task of teaching easier. When she was questioned about what she doesn't like teaching, she referred to human physiology and biochemistry, considering the latter more complicated for students because it is very abstract and involves chemical knowledge.

Reporting on her classes on evolutionary theories, she commented that from the first moment she makes clear that she believes in God and believes in science, squelching the notion that believing in science does not mean not believing in God. Liliane showed in her speech that she sees no problem in believing in divine creation. The teacher begins her classes explaining that everything happens with the permission of God, although she makes clear in her classes that she will not discuss what is right or wrong in religious matters. "There was a time when people would believe or in the evolutionist theory or in the creationist theory, but creationism is not only about God creating things, period. And that they do not change, He created things as they are. The other theory does not speak at any moment of divinity except that species evolve."

About her teaching methodology, she reports that in the first classes she does not like to answer the students' questions because she believes that the students need a previous explanation. She comments on creationism, posing it as theory and comparing it with evolutionism. She presents that God created all things and things evolve, bringing together science and God. She explains things gradually, first creationism, followed by Lamarkism and Darwinism, and later introduces concepts such as speciation and races. According to Liliane, the students will understand naturally, she does not comment on the man coming from the monkey.

When she was asked about the conflict between God and science, she answered that she feels comfortable about it and has no difficulty dealing with it, because she had a religious background that allows her to talk about it freely. She emphasized that she does not like to work on the content of evolution in the preparatory courses for admission to higher education because she does not have time to explain and discuss it better with the students, since she has to be more succinct and speak directly about the theories due to the scarcity of time. When she was questioned about the students' understanding of this subject, she responds that they have no difficulty in understanding it, she has never seen any problems, since she started the class by solving conflicts and showing that she believes in God. As she summarized, "One thing does not exclude the other."

Teacher Sheila: "There is a need to separate religiosity and science, the difficult thing is not to understand the theory, but the separation."

Teacher Sheila is forty years old, married, has two daughters, claims to be Catholic and has fifteen years of teaching experience. In biology teaching, she reports that the contents she likes the most are: ecology and human physiology, emphasizing that she also likes to teach sex education. The content that she has less affinity with is biochemistry, saying it is difficult to give a class about cellular metabolism because it is an abstract content and that the students have difficulty with it. She mentioned that she considers cell division equally complex for students to understand.

About classroom work, she mentioned that she approaches the topic of evolution in a very tranquil manner, stating that she believes in God and that it was difficult in college to separate the two. Sheila realizes that there is a need to separate: religiosity and science. The teacher, during the interview, was concerned about showing her development in understanding the subject, stating that the way she was presented at the undergraduate course was difficult at first, but later, there was a personal elaboration to separate religiosity and science. She perceives the conflict between Genesis of the bible and the Big Bang; and says that separating them was difficult for her.

She explains to students about their process of understanding scientific thinking, their doubts and conflicts. However, she stated that she does not have many conflicts with the students; they do not question much, but they always ask if she believes in God. She said once she had an argument with a Protestant student. The student said she does not believe in theories contrary to creationism, she reported that she spent half an hour arguing with the student about the subject, insisting that the student had to know about evolutionism. She finished the discussion by telling the student that one day he would need to know about this content.

Teacher Amanda: "I do not try to convince them that those theories are absolutely right, and that creationism is totally wrong, I respect the religiosity of each one."

Teacher Amanda is thirty-eight years old, married, and has one child. She has fourteen years of teaching experience and when she was asked about religiosity, she says she does not follow any religion. The content she likes to work on is genetics, zoology, and human physiology.

She does not feel comfortable with botany because she believes there is a certain distance between the content and the student's reality, there are many concepts and complex nomenclatures. Regarding her approach in the classroom to teach evolutionary theories, she stated that she seeks to explain the contents in a more realistic possible way, bringing clues from the daily life of the student and she also uses audio visual tools, slides, and posters to facilitate students' understanding.

She didn't talk about the religious question, or even talk about God, she is concerned about presenting only theories related to the school content, she only cites creationism, believing that it is part of the evolutionary history in scientific knowledge. "I have not had much resistance from the students, I do not try to convince the students that those theories are absolutely right and that creationism is totally wrong, I respect the religiosity of each one, I present the content, theorists and theories, without trying to force them to accept that as an absolute truth, I had no problem with that."

Teacher Mariana: "God's time is not our time."

Teacher Mariana is thirty-two, married, has two children, and twelve years of experience. She claims to be Catholic, but she attends the Spiritist center. She likes to teach human anatomy, zoology, histology, and genetics. The content she doesn't like much are: cell division, biochemistry, embryology, and botany. Her classes on evolutionary theories are based on newspaper news, fossil finds, games, jokes, book excerpts. She claims she loves the books of Richard Dawkins, an ethologist, evolutionary biologist and British author, author of the famous book "*The God Delusion*." She likes to begin class by reading to the students excerpts from the Bible, beginning with the book of Genesis (The Origin of the World and Humanity). "Adam made love to his wife Eve, and she became pregnant and gave birth to Cain. She said, 'With the help of the Lord I have brought forth a man.' Later she gave birth to his brother Abel." (Genesis, 4:1–2). Mariana said that she also cites the part in which Cain joined his wife, who conceived and gave birth to Enoch. Cain built a city and gave the city the name of his son Enoch. In this context she problematizes with the students about the incoherence in affirming that Cain united with his wife, since there were no other women only Eve, who was his mother. Consequently Cain would have committed incest. Mariana asks, "Is the Bible permitting sexual relationships between children and parents that in my view is not correct?"

Another question raised by the teacher is about Adam, the first man on earth, created, according to the Bible in the image and likeness of God and if he had a navel, confronting the biological explanation with the religious. Biologically, the navel is just the scar left by the umbilical cord, there is no connection with our organs. If Adam was made by God without being raised in a woman's womb, it does not justify having a belly button. She also suggests that protestant students ask their pastors about these inconsistencies and doubts. Mariana claims she believes in God and feels the need to unite and give real meaning to creationism and evolutionism in a way that makes sense. Thus, the relation between scientific knowledge and creationist theory is summed up by the teacher trying to solve the conflicts from the following reasoning: God is the creator of everything and created the planet according to the evolutionary theory, from the Bing Bang. She explains to the students that the biblical part cannot be taken literally, it must be interpreted, because God's time is not equal to the time of men.

Mariana says that the less receptive students are Protestants, especially when some biblical passages are analyzed to make some comparisons. However, she asserts the conflict can be solved by placing God as the originator of the Big Bang and presents her own theory based on personal contact with university colleagues and several biblical passages. She managed to unite the theories of creationism and evolutionism; the bible cannot be understood literally, there is a need for interpretation. She tells the students that Adam was not a person but a village and that after a possible disagreement among the people of this population there was a separation and the village of Eve was created and from these two initial settlements with migrations others appeared. Mariana stated,

> From the two other settlements arose from the settlement of Cain and the settlement of Abel, so I mean, it is not a person Adam, Eve, Cain then Abel. They are several people forming towns with these names and when he says that Cain killed Abel, it is a group which exterminated the other, in order to justify and even find an acceptable explanation for when he says that Cain made love to his wife, then other populations popped up other populations, you see? So I explain to them like this, I do not want to stop believing in God, I believe in God, but I do not want to say what is in the Bible is a lie, although there are some things that we keep questioning a lot, I try to explain it in this way, that the world was created by God, but why could not have been God that generated the Big Bang?

Teacher José: "I defend scientifically and rigorously the theory of biological evolution."

Teacher José, who is fifty years old and married, has two daughters, and has twenty-four years of experience. Jose practiced dentistry for many years, then took a biology course. Teaching is his true passion. He grew up in a Catholic family, but he defined himself as an agnostic. He likes to teach the contents of genetics and biological evolution, he thinks that systematics and classification contents difficult to be approached. In his classes on evolutionism, he speaks to students that creationism exists, but he makes it very clear that this theory is not related to the study of science. The basis of any study of biology is the theory of evolution defended by Darwin. He does not find it appropriate to create conflict during class by discussing religious matters. He then commits to teach the theory of evolution, explains all its basis and how important it is, and that it is the foundation of the study of all biology. He does not clash with students who try to bring the discussion to the religious side, making it clear that the study will be about biological evolution.

DISCUSSION AND IMPLICATIONS

When we started the interviews, we did not imagine they would be so rich in helping us in our understanding of how biology teachers comprehend and work on evolutionary theories in the classroom. Undoubtedly, this is a theme that still needs to mature, to be treated with seriousness in academic circles. From popular culture in Brazil, we are led to believe that religion is not discussed nor problematized, as if religious principles and formulations could coexist harmoniously with scientific hypotheses and tests. With the appropriation of scientific knowledge by an increasing number of people in the population and with the growing number of people having access to schooling, tensions and conflicts generated by divergent ideas become ever larger.

When we started this research, we thought we would only present data on how evolutionary theories were taught in Brazilian schools and that creationism was only an allegory in the formation of our students. Through interviews, however, we realized that this subject goes far beyond the classroom or the training of teachers; it is a subject of cognitive and intellectual development that goes beyond scientific training and is rooted in the personal development of each one of us. Brazil is a country where religion is presented as a strong component in the formation of people, and religious participation is attributed to the formation of values and practices of human valorization. Thus, it is rare to see Brazilian people assuming they do not believe in God or that they are not bound to any religion.

An interesting point about this research was also being able to share it with other researchers. We realized that other Brazilian researchers have already done research with the theme and had different ways of approaching the problematic of evolutionary theories in teaching practice. The researchers Teixeira and Andrade (2014) published the article, "Entre crenças pessoais e a formação acadêmica: como professores de biologia que professam fé religiosa ensinam evolução?" With this questioning, the researchers interviewed high school biology teachers with the main question of their interview being: Why do I teach evolution? The researchers had a hypothesis of an identity crisis in relation to faith and science. The research and the reports produced did not let them affirm that this crisis of identity exists—but together with our observations, they emphasize the importance of the theme and the search for answers that favor and strengthen the scientific knowledge.

Another group of researchers that drew our attention were Silva, Silva, Souza, and Araújo (2013), who together dedicated themselves to a similar research carried out in England in 2009 under the title of "Rescuing Darwin." The researchers confronted the interviewees with the statement: "Atheistic evolutionism argues that evolution makes it unnecessary and absurd to believe in God." The researchers were able to perceive that Brazilians tend to consider the affirmation to be false in a greater percentage than the English do, showing the strength of religious beliefs in the Brazilian population.

We also find research carried out by Sanches and Danilas (2012) that proposes a harmony between religion and science in Brazil. This research was carried out by theologians who based the relation in the perspective of the Christian tradition, emphasizing that it was in the Western world, on the bases of the tradition, scientific and technological issues have developed. By the interviews, it became evident that biology teachers working on evolutionary theories are confronted with their own beliefs and positions regarding God and science, we realize that from the six teachers interviewed only one really had already a definite position on this issue. Another teacher says that she went through a process of personal transformation, reflecting critically on her beliefs and certainties in relation to the scientific world that was immersed from her undergraduate degree in biology. The other teachers mentioned resolving the conflict between God and science by expanding their belief to the scientific world, that is to say to themselves and to the students that God is in everything and allows everything, including understanding that evolutionary theories exist only with divine permission.

After listening to interviews and transcribing them, we applied content analysis, that according to Bardin (2016) allows us to identify recurrent themes that were evident in the interviews. From this, we begin to group the ideas in the teachers' discourse. A recurring unit in the speech was the God

Table 4.1.

Name	*Identification Information*	*Subject with More Affinity*	*Subject with Less Affinity*	*Dimensions of Analysis*
ALICE	38 years old Experience: 19 years Married with 2 children Catholic	Genetics	Botany	Associates God and Science
LILIANE	43 years old Experience: 15 years Married with 1 child Spiritist	Genetics	Human Physiology and Biochemistry	Associates God and Science
SHEILA	40 years old Experience: 15 years Married with 2 daughters Catholic	Ecology, Human Physiology and Sex Education	Biochemistry and Cell Metabolism	Separates Science from religiosity
AMANDA	38 years Experience: 14 years Married with 1 child Does not follow any religion	Genetics, Zoology and Human Physiology	Botany	Separates Science of Religious
MARIANA	32 years old Experience: 12 years Married with 2 daughters Catholic	Human Anatomy, Zoology, Histology and Genetics	Cell Division, Biochemistry, Embryology and Botany	Associates God and Science
JOSE	50 years Experience: 24 years Married with 2 daughters Agnostic	Genetics and Evolution	Systematics and Classification	Separates Science from religiosity

vs. Science question. Although some teachers did not speak literally on this issue, they all mentioned their strategies for approaching evolutionary theories, some perceiving this dilemma as solved, others mentioning this dilemma as a process for understanding the subject. Thus, we conclude that from this dilemma the statements of the teachers in the resolution that each one can present at this moment to the dilemma: God vs. Science. From diverse studies and positioning, we noted the following perspectives of thought. In order to have an overview of the interviewees and the dimensions of analysis, we produced a summary table (see Table 4.1).

Perspective 1—Links God with Science

We perceive that most teachers are women, who describe that the conflict does not exist but at the same time they speak to students about believing in God and that He is the creator of all things. This is a contradictory dimension in the teachers' speech, though their discourse is coherent and scientifically acceptable. This was also the dimension most presented by the teachers, with perspectives built from their belief in a single Judeo-Christian God. In the case of the teachers interviewed, based on their Catholic, Protestant, or Spiritist religion. The cognitive movement of this dimension is to generalize individual and religious understanding to the scientific milieu, presenting for the students creationism not as a mythological or legendary allegory, but as a possible and viable theory of being accepted in the scientific world.

In this perspective we had the speeches of the teachers: Alice, Liliana, and Mariana. Alice said: "In relation to God I approach science, because for me they are the same thing. I do not confront the teachings of the catholic church with scientific evidences: God and Science are the same thing with different explanations. God may have created natural selection and evolution. Science shows the same context of God only in a more scientific way." Liliane said, "I begin by saying that I believe in God and science. One thing does not exclude the other." Mariana said, "God is the creator of everything and created the planet according to the evolutionary theory, from the Big Bang."

In the attempt to approach God and science, the statement that most attracted our attention was from Mariana, because in her effort to make this approach meaningful, she constructed a theoretical alternative by affirming that this is her interpretation of the Bible. In her account, there is a partial break with religious logic that is reestablished with her interpretive efforts. The testimony of the three teachers grouped together in this dimension presents the belief that God, in the Christian perspective, created and allowed everything including evolutionism. The narratives present contradictions and at times they do not seem to come from people connected to science. The contradictions and inconsistencies are not perceived by the three teachers, for they are products of their critical and scientific reasoning.

We also thought that the fact that teachers approach evolutionary theories would be enough to minimize the creationist approach in school. However, it was evident that introducing evolutionary theories along with the belief in a God who creates all things, omnipresent and omnipotent generates, at least, a tendency in the possibility of creationism to be really the origin of everything, that is, to be above the evolutionist theory.

Another aspect that seemed important to us, was the absence of different creationist approaches, we know that the Judeo-Christian biblical approach is only one of the possible explanations for the divine creation of the universe and of all beings. Each culture, over time, has woven a divine explanation for the complexity of the emergence of life. African, indigenous, Egyptian, Aztec, Mayan, Indus, and Eastern cultures, each of these peoples in the course of human history, have produced legends, or mythological, creationist explanations. The teachers interviewed only referred to Christian creationism as the only possibility, based on a monotheistic religion, leaving aside other explanations or understandings produced in the course of Humanity, coming to be affirmed in one of the statements that "God and science are the same thing."

Perspective 2—Separates Science and Religiosity

We perceive that the conflict between God and science has been experienced, or is still in the process, tending toward scientific understanding above religious. For this, we put first the word science and then the word religiosity (and not God). The reports of the teachers grouped in this perspective seemed to us more flexible and very much based on a personal development, from a critical reflection on their beliefs and thoughts. This dimension was less contradictory than the first. Teachers are aware of the existing conflict and seek ways to ease the discrepancies arising from religious understanding. Although three of the six statements are grouped in this perspective, only one of the teachers referred to anything other than a specific religion, assuming an agnostic.

For example, Jose stated, "I defend scientifically and rigorously the theory of biological evolution." Amanda said, "I have not had much resistance from the students, I do not try to convince the students that those theories are absolutely right and that creationism is totally wrong, I respect the religiosity of each one, I present the content, theorists and theories, without trying to force them to accept that as absolute truth, I had no problem with that."

In the statements grouped in this perspective we have three different perspectives: Teacher Sheila, who shows students that the understanding of evolution requires a paradigm shift, with the abandonment or suspension of some religious beliefs so that the separation of science and religiosity can occur; teacher Amanda who seems to defend an impartiality and the teacher José that presents a firm position with respect to the scientific thought and the evolutionary theories.

FINAL CONSIDERATIONS

This research demonstrates how the teaching of evolutionary theories is still a controversial and delicate subject in teaching practice. The "God X Science" dilemma was perceived with intensity in the interviews and from this, the way of solving it led us to group the statements in two perspectives of analysis: the association of God and science and the separation of God and science.

From the fact that we initially understand that all biology teachers are scientists and are committed to studying and disseminating scientific knowledge in their field, we were very struck by the testimonies that link God and science and even more with those who seek to connect creationism and evolutionism in the same thought process.

We realize that the teachers who have evidenced this perspective have a great concern in creating conditions that do not discredit creationism, maintaining it as a possible theory for the creation of man and the universe. Three in six interviewed teachers are in this perspective, which leads us to believe that this way of solving the "God X Science" dilemma still finds supporters that tend toward a divine explanation, even with academic formation and the practice of teaching of science.

Only one of the testimonies showed us a path to the development of scientific thought. Teacher Sheila told us about her personal development to understand scientific issues and mentioned how difficult it was to trace this path during her academic background. From her personal experience, she attempts to show the same path for students by encouraging them to break with personal beliefs for understanding scientific concepts.

Professor Sheila, along with two other teachers interviewed solve the dilemma, separating religiosity and science. In this perspective, only one of the teachers had a firm position in relation to his role of professor of biology, affirming that he defends scientifically and rigorously the theory of the biological evolution. In that speech, we perceived the difficulties of the teachers who propose to defend the evolutionary theories in the classroom, with colleagues, and in the school community.

In short, the research presented to us the complexity and the unfolding that the teaching of evolutionary theories contain in the teaching practice. Besides the didactic book and the academic scientific formation, the evolutionary theories pass through the break of paradigms that in the twenty-first century still are present in the speeches and teacher practices.

REFERENCES

Amabis, J. M., and Martho, G. R. (2014). *Biologia das populações.*

Bardin, Lawrence. (2016). *Análise de conteúdo.* Lisboa: Edições 70.

Belei, R. A., Gimeniz-Paschoal, S. R., Nascimento, E. N., and Matsumoto, P. H. V. R. (2008). O uso de entrevista, observação e videogravação em pesquisa qualitativa. *Cadernos de educação 30*(1): 187–199.

Bourdieu, P., Curto, D. R., Domingos, N., and Jerónimo, M. B. (1989). O poder simbólico.

César, S. J., Sezar, S., and Caldini, N. (2013). Biologia. *vol. único-Ed. Saraiva.*

Cicillini, G. A. (1997). A produção do conhecimento biológico no contexto da cultura escolar do ensino médio: a teoria da evolução como exemplo.

Darwin, C. (2009). A origem das espécies.

Dawkins, R. (2017). *O gene egoísta.* Editora Companhia das Letras.

Erbs, Rita T. C. and Abrahão, Maria Helena Menna Barreto (2018). Memoriais de Formação—A escrita e a escuta no início do curso de Pedagogia. In Abrahão, Maria Helena Menna Barreto (Ed.), *A nova aventura (auto) biográfica: TOMO II.* Porto Alegra, Brasil: Editora PUCRS.

Herculano-Houzel, S. (2017). *A vantagem humana: Como nosso cérebro se tornou superpoderoso.* Editora Companhia das Letras.

Herculano, H. S. (2002). O cérebro nosso de cada dia: Descobertas da neurociência sobre a vida cotidiana.

Herculano-Houzel, S., and Kaas, J. H. (2011). Gorilla and orangutan brains conform to the primate cellular scaling rules: implications for human evolution. *Brain, Behavior, and Evolution 77*(1): 33–44.

Lopes, S. G. B. C. (2015). *Bio.* Saraiva.

Oleques[1], L. C., Bartholomei-Santos[1], M. L., and Boer, N. (2011). Evolução biológica: percepções de professores de biologia. *Revista Electrónica de Enseñanza de las Ciencias 10*(2): 243–263.

Sanches, Mário A., and Danilas, Sergio. (2012) Busca de harmonia entre religião e ciências no Brasil: reflexões a partir do ano de Darwin. Revista Teocomunicação. Porto Alegre, 98–118.

Silva, Heslley M; Silva, Paloma R.; Souza, Ana C. e Araújo, Elaine S. (2013). A influência da religiosidade na aceitação do evolucionismo: Um estudo em uma amostra da população brasileira. *Revista Conexão. UNIFOR-MG, Formiga 8*: 1–19.

Teixeira, Pedro and Andrade Marcelo. (2014). Entre as crenças pessoais e a formação acadêmica: como professores de biologia que professam a fé religiosa ensinam evolução. *Revista Ciência e Educação, Bauru 20*(2): 297–313.

Chapter Five

Pedagogical Strategies for the Problem of Teleology in the Teaching of Biological Evolution

Beatriz Ceschim and
Ana Maria de Andrade Caldeira

Alternative conceptions about biological evolution (such as teleological thinking) may hinder the learning process of scientific consensus, so it is very important that the process of instruction is based on such alternative conceptions to aim at the conceptual change of students. In the present work, pedagogical strategies were elaborated that include a conceptual network, aiming to help the nonteleological understanding of evolution for teachers' information. It was the objective of the research to investigate the effectiveness of such pedagogical strategies elaborated.

It is common in the literature to refer to conceptions that are configured differently from the Darwinian conceptions of the evolutionary process, both in basic and higher education (e.g., Sepúlveda and El-Hani 2014; Sepúlveda, Mortimer, and El-Hani 2013; Gregory 2009; Gregory and Ellis 2009). It is a common conception that biological evolution is the process whereby the body produces a response to the needs of a given time (i.e., that evolutionary change is triggered by need and that the individual transformation of organisms results in adaptation to the environment) (Brumby 1979; Bardapurkar 2008). It is an explanation with a teleological structure because it uses the final state (or goal) to explain the evolutionary why.

Studies carried out in Brazil show that even teachers who work in basic education have teleological conceptions, mainly because of the influence that creationist thinking can have on explanations of biological knowledge (Silva, Andrade, and Caldeira 2014). Especially before Darwin, assertions that include "in order to" explain the existence of a biological structure or process pointing to the actions of conscious entities, such as God. Thus, when it is said that the heart beats "in order to" circulate the blood, it is a shortening of a phrase like "it was God's intention that the blood circulate through the body of vertebrates and he knew that the beats of a heart were a good way to do that. "In other words, for each one "in order to," one makes an affirmation concerning the omniscience and omnipotence of God, especially with regard to his capacity for foresight (Rosenberg and Mcshea 2008). This would be an explication of a teleological explanatory structure committed to theological visions.

Teleology is an intuitive form of understanding closely associated with intentional explanations of purposive behavior (Lombrozo and Carey 2006) as are reasonings created which establish a relation of analogy to the desires and capacities of foresight possessed by humans (or God). It consists of inadequately generalizing from the notion of thinking organisms (humans) creating artifacts. According to Kelemen (2012), the legacy of teleological thought formed in childhood perpetuates a generalization of thinking that results in difficulties understanding the evolutionary process.

As Ferrari and Chi (1998) note, the transposition of the metaphor from artifact design to the origin of living is inadvisable because ontological characteristics that are not shared by artifacts and living beings can be erroneously attributed to the evolutionary process. In interpreting the origin of a knife, for example, the effect of this object is obviously correlated with the necessity that motivated its origin: the knife was created for the purpose of cutting, so the need for an object that cuts led to the creation of the knife. The essential difference between artifacts and living organisms is the presence/absence of intentional foresight whereby a project is formed with the intent for it to exhibit a given performance (Dawkins 2001).

Therefore, because the formation of new inheritable structures and functions are the result of contingent processes and not due to organisms' needs, it is necessary to consider conceptions that point to a kind of anticipation of the process of origin of evolutionary innovation as alternative conceptions, evidently incompatible with the scientific knowledge. It is therefore assumed that although the student may use teleological thinking to interpret other aspects of nature (such as deliberate human behaviors and the creation of artifacts) or to interpret the purpose of his own life or the existence of the world (which may be connected to the student's beliefs and religions), there is a need for conceptual change to take place specifically with regard to the necessary rejection of teleology at the origin of evolutionary innovation. The perspective of conceptual change adopted in the research is that of Vosniadou (2007, 2008), characterized by the need for instruction for an extended and continuous time (beyond an instant substitution of conceptions), intentional on the part of the students of passive direct teaching) and that contemplates moments for the expression of the students (beyond the instruction accomplished through the hegemony of the expository model).

METHODOLOGY

The methodology used here is qualitative (Yin 2016), and data collection included audio recordings of nine meetings of two hours each along with the application of an individual written questionnaire. Data analysis was performed by devising categories based on content analysis (Bardin 2013), which allowed the classification of participants' conceptions in four major categories about the evolutionary perspective: (1) Synchronic analytical perspective (does not recognize the need for historical and evolutionary explanations); (2) Transformational perspective (emergence and loss of characteristics according to the need of the organism in an essentialist perspective; use and disuse mediated by law of acquired characters); (3) Selective perspective (one characteristic is determined after a selection regime); (4) Process pluralism (considers natural selection as well as nonadaptive hypotheses to explain biological traits); and in two broad categories concerning teleological thinking: T1. The evolutionary process is endowed with anticipation and meets goals and T2. The evolutionary process is progressive.

The meetings were held at Paulista State University "Júlio de Mesquita Filho" with eleven biology teachers in training (A1–A11) who were completing a university degree in evolution and had already studied other subjects using evolutionary approaches.

The research can be subdivided into three main stages: (1) survey of previous conceptions, reading, and discussion about the basic concepts of evolutionary biology (first to third encounter); (2) reading and discussion concerning evolutionary developmental biology—or evo-devo—and genetic drift (fourth encounter); and (3) reading and discussion concerning the use of teleology in biology (fifth-to-last meeting).

At all stages, trainee teachers were continually invited to express their own understanding so that it was possible to question the compatibility between the ideas expressed by the participants and the content of the materials studied. Strategies used in the pedagogical interventions (direct instruction and discussions) for the purpose of this research are briefly mentioned below.

Strategy 1—Imperfections of Living Things

We discussed the presence of structures that do not increase an organism's fitness, the so-called "imperfections" of Futuyma (2009) and Reece et al. (2015), such as (1) dandelion flowers that present attractive petals for pollinators but have sterile pollen and develop seeds without fertilization; (2) the human appendix; (3) nipples in men; and (4) the connection between the respiratory and digestive systems that may causing asphyxiation by food. The aim was to illustrate that characteristics that exhibit an irrelevant or faulty operation may be preserved phylogenetically (to highlight the absence of teleology and progress to perfection in evolution).

Strategy 2—Population Thinking

We introduced population thinking with the example of the black widow spider (*Latrodectus hasselti*). In this species, males that are cannibalized during copulation fertilize approximately twice as many eggs. One possible explanation is that copulation wherein the male submits to cannibalism by the female has a very long duration (compared to copulation in which the male resists) because the female is "distracted," feeding on the male. The hypothesis would be that if the behavior of submitting to cannibalism is genetic, offspring should also exhibit this behavior. As the male suffers the loss of the structures necessary for fertilization and may not survive copulation, the first fertilization in which a male participates could also be his last, regardless of the cannibalism of the female. In this scenario, it is much more advantageous for the male to produce many eggs in this single fertilization, submitting to cannibalism (Meyer and El-Hani 2005). With this example, we aimed to introduce the concepts of differential reproduction, variability, and heredity to support natural selection discussions.

Strategy 3—Role of Pleiotropy

We discussed the implications of the existence of pleiotropic genes to provide an alternative interpretation of the presence of structures exhibited in organisms because such genes can lead to the perpetuation of characters not selected directly by natural selection.

Strategy 4—Characterization of Natural Selection

We discussed the different types of selection results: "Despite elimination, there will always be abundant variation, consisting not only of those truly excellent individuals but also, indeed, of all those who are not so bad as to need to be removed by elimination" (Mayr 2004, 152). The defense that natural selection is not able to positively select only the "best" demonstrates that natural selection is not capable of engendering "perfect" results in the distribution of organisms in populations.

Strategy 5—Sexual Selection vs. Natural Selection

We discussed the evolutionary effects of fixing onerous characteristics through sexual selection, which may not be very advantageous in terms of survival, including the cases of sexual selection with regard to (1) the peacock, where the choice of the female is influenced by the appearance and behavior of the male; (2) amphibians, where the male attracts the female with long or short calls, and the female usually prefers the longer call (Campbell and Reece 2010); and (3) fish, where males attract females by exhibiting a colorful and bright appearance (Campbell and Reece 2010). While a certain characteristic may help the male attract a partner, the same characteristic can increase the risk of predation, which shows the lack of progress or teleology in the evolutionary process.

Strategy 6—Characterization of the Origin of Variability

We were able to subdivide the process of evolution by natural selection into two stages: the phase in which variations appear and the phase of natural selection itself. Heritable mutations (those occurring in germ cell DNA) were differentiated from nonheritable mutations (those occurring in somatic cells), allowing discussion of the rejection of the inheritance of acquired characteristics.

Strategy 7—Genetic Drift

The example of Lençois Island (located approximately thirty kilometers from Maranhão, Brazil) was presented, in which there is a proportion of albinos well above the proportion in other human populations. It is an example of genetic drift, as the proportion of the island's albinos is explained by the founder effect: albino families colonized the island and had offspring with reproduction among relatives. The island has a lot of solar exposure, and albinos often suffer from skin cancer. Thus, the fact that the proportion of albinos is higher on this island cannot be explained by natural selection. The purpose of this discussion was to emphasize that (1) evolution does not always happen by natural selection and (2) evolution does not always lead to better living conditions—again with the aim of highlighting the absence of teleology in evolution.

Strategy 8—Embryological Constraints and Evo-Devo (Evolutionary Developmental Biology)

We presented an analogy that posits a faucet through which spheres of different diameters pass (Meyer and El-Hani 2009). The faucet, which illustrates the source of variation, can let spheres pass through without the permeability of a sieve (the sieve illustrates natural selection), but the faucet may be fully open or partially closed (illustrating the embryological constraints on change). When the tap is partially closed and there is no sieve, the distribution of spheres is determined only by the tap. However, when the tap is fully opened and there is a sieve, the distribution is determined by the sieve. When the tap is partially closed and there is a sieve, the distribution results from both tap closure and sieve presence. The use of the analogy made it possible to show that the origin of variability undergoes embryological restrictions, altering the phenotypic distribution of living beings. The aim was to decharacterize the origin of variability as infinitely plastic or teleological and to emphasize that the distribution of living traits is also a consequence of the contingencies of development.

Strategy 9—Phenotypic Innovation and Developmental Regulatory Genes

We addressed the mutations that can occur in the genes regulating development as a source of variation. Carroll (2006) notes that variation in developmental genes may explain evolutionary leaps, which helps trainee teachers understand the source of nongradual variability as contingent.

Strategy 10—Distinction between Innate Behavior and Deliberate Behavior

The distinction between innate and deliberate behavior is required in discussions concerning teleology because the deliberate behavior of thinking organisms is undoubtedly a teleological behavior. However, the generalization of deliberate behavior to the understanding of innate behavior, for example, can be a misconception that supports the improper use of teleological biology.

Strategy 11—Teleology and Natural Selection

When the students structured selective narratives, we directed efforts to identifying the hypothetical character of the participants' explanations because apparently such narratives were understood as satisfactory elucidation that did not demand further verification. We introduced Caponi's (2012) concept of function and conducted a discussion mediated by an analogy: the gravitational influence of the moon on the tides. We emphasized that for the moon, the effect exerted on the tides is not the reason for its existence. However, in interpreting living traits, it is common to explain the presence of traits based on the effect (understood as the function) produced by the traits, which is assumed to be a trait that was positively selected in the past. The intention was that, if participants argued that the moon did not arise to have an effect on the tides, we would use the same argument to emphasize the absence of teleology in the origin of variations, for example, "mutation did not arise because of an organism's need."

We used the example of the bad smell produced by the skunk (Caponi 2012) to illustrate the hypothetical character of the narratives and "stories" that were proposed by the participants when trying to explain the evolutionary causes of the presence of a characteristic. The participants were presented with three situations in which skunks would produce the bad smell: (a) skunks produce a bad smell, which keeps away predators; (b) skunks produce the odour as a consequence of excreting substances toxic to the organism; and (c) skunks do not produce a bad smell when they do not eat foods containing toxins requiring elimination. The three situations were explored in isolation by the participants, who attributed functions to the bad smell as well as an explanation of ultimate cause to this characteristic.

In a second stage, the three situations were presented simultaneously: the skunk produces a bad smell due to the need to eliminate toxic excreta, and the smell can ward off predators, but in an alternative diet, the skunk does not eat toxic substances, and the elimination that causes odors is not necessary. After describing this more complex picture, the participants were asked to assign a function to the bad smell and produce hypotheses regarding its evolutionary causes. The purpose of this activity was to illustrate how the target of the selection can be different from the target of the functional assignment because, if in the example there are two advantages exhibited by the characteristic "bad smell," which are elimination of toxic excreta and protection against predators, why was this feature selected positively after all? We did not seek to reach a final answer but rather to produce a topic of discussion that would allow reflection on the limitations of assigning selection narratives without empirical data.

Strategy 12—Synthesis of the Teleology-Evolution Relationship

By discussing the appropriateness of employing teleology in the interpretation of artifacts and the inappropriateness of interpreting the origin of living traits, the trainee teachers performed a rereading of old inadequate assumptions that presuppose direction, progress, or appearance of variations according to the needs of an organism. At this point, therefore, participants conceptualized the nature of evolutionary mechanisms (gene mutations, natural selection, genetic drift, and developmental constraints/mutations) and described the inappropriateness of teleology in their interpretation.

Considering that teleological thinking may be associated with beliefs and religion and that it is permissible in the biology of near causes (or functional biology) (Toepfer 2012; Mossio and Bich 2017), the change in the conceptual profile should not aim at the radical substitution of a previous form of thinking for scientific knowledge, but mainly the awareness of what are the appropriate and inadequate uses of teleology to interpret different phenomena. Thus, it can allow an understanding of the different roles of teleology that are not in accordance with the scientific knowledge.

RESULTS

The research carried out allows us to conclude that (a) the trainee teachers initially presented teleological conceptions, which hindered their understanding of the evolutionary process (see table 5.1); (b) the strategies of the pedagogical interventions allowed the formulation of nonteleological explanations for evolutionary processes; (c) the final conceptions were closer to the scientific model of understanding biological evolution; (d) for some participants understanding evolution as a biological process expanded beyond a teleological selectionism; and (e) the selection of conceptual questions used in the strategies contributed to promote possible ways for the conceptual change concerning the problem of teleology in teaching and learning of biological evolution.

Table 5.1 includes the conceptions of the trainee teachers (A1–A11) classified in conceptual profile zones represented by numbers 1 to 4 as well as the classifications of teleology (T1 and T2).

In general, it is possible to affirm that in the conceptual evolution of the trainee teachers, there was a decrease in the frequency of transformational perspectives and teleological formulations, increasing frequency of selective perspectives and, to a lesser extent, increasing pluralistic perspectives.

Did the Trainee Teachers Reach a Conceptual Change?

With regard to the conceptual profile, we observed only one conception indicator located in the first conceptual zone, "synchronic analytical perspective." The conceptions located in zone 2, "transformational perspective," were very frequent in the first meetings and less frequent at the end (eight times in total). The conceptions grouped in zone 3, "selective perspective," were present from the first stage but became more frequent throughout the development of the group, especially at the end (twenty-seven times in total). Zone 4, "process pluralism," became more frequent after the second and third stages, being absent in the first stage (ten times in total), as shown in table 5.1.

Table 5.1. Distribution of research subjects' individual conceptions. When there is no "T1" or "T2," it means that there was no teleological formulation in that step. The - symbol indicates that the subject did not participate in the step. The * symbol indicates that the subject was present but did not issue statements during the step.

Subjects	Stage 1	Stage 2	Stage 3	Questionnaire
A1	1	*	2	3
	T1		3	
	T2			
A2	2	3	–	–
	3	4		
	T1			
A3	3	2	3	3
	T1	3		4
		4		
A4	T1	*	3	3
	T2			
A5	2	3	4	3
	3	T1		4
A6	3	3	3	3
			4	4
A7	2	3	3	3
	3			4
A8	3	–	–	–
	4			
A9	2	*	*	3
	3			
	T1			
A10	T1	3	4	3
	2			
	3			
A11	2	–	–	
	T1			

It can be seen, according to table 5.1, that conceptions located in zone 2 (transformational perspective) appeared frequently in the first stage but declined in the following stages, until they disappeared completely in the final questionnaire. The conceptions of the selective perspective (zone 3) were more frequent in the first phase and in the final questionnaire. The pluralistic perspective (zone 4) increased in frequency from stage 2, due to the introduction of discussions about evo-devo, genetic drift and the phenotypic effect of pleiotropic genes at this stage but did not reach the same frequency as the selective perspective in the final stages.

Characterization of Teleological Conceptions

Referencing the nine meetings with the trainee teachers, in the beginning we obtained more indicators of teleological conceptions, though these became rarer in the final meetings and absent in the final questionnaire. Teleology was present in conceptions based on Lamarckian premises, as in selective premises, as observed through indicators present in the participants' conceptions, as in such examples of transcribed speech:

> *Participant A3:* Regardless of the mechanism that occurs, an hour will have to be selected. In this example of the island, I know that if thousands or millions years occur, the number of albinos will not increase, and it will not happen again as it has previously. The number of albinos will have to be modified as it will be smaller the current number of albinos [The student argues that it is impossible for albinos to survive in a large number over time on an island with a high incidence of sunshine—cf. other information on this example of albinos in Strategy 7 earlier].

Based on the defense of such a statement, it is possible to note that teleology is not only linked to Lamarckian (or classical Darwinian) impressions of explanations for evolution (in the sense of understanding that the use or disuse and the law of acquired characteristics are plausible explanations) because even in a framework that is compatible with the Modern Synthesis, which allows for natural selection and gene mutations, it is possible to produce teleological pictures that show an understanding of the evolutionary process. As the above discourse has shown, it is possible to understand (through a mistaken conception) that natural selection leads to predictable results, which have outcomes that are necessarily served (such as the unavoidable elimination of albinos on Lençóis Island).

Progressive teleology, according to some participants, can act as a way of conceiving the world (which may be related to the beliefs and religion of the trainee teachers), which is transposed by the student in order to understand biological evolution. Another common form of teleology is that relating to the attendance of needs, which is the conception based on the supposed presence in the organic structure of a capacity for self-awareness and anticipation to respond to immediate needs through the production of heritable evolutionary innovation.

> *Participant A7:* Consider the wisdom tooth. Some people are still born with wisdom teeth, but some people are no longer born with wisdom teeth because they are unnecessary. Foods used to be harder, people needed to chew longer, and they needed more teeth to chew with. Currently, this need no longer exists for humans.

Participant A7: You can talk about the fontanel, too. Formerly, the skull was much larger, but it currently has a tendency to decrease its formation because children are no longer being born by normal birth. Instead, most children are being born by cesarean section, so there is no longer a need for the larger fontanel. The tendency is that children tend to have a decreased fontanel.

Participant A10: At some point, the antibiotic will kill the bacteria that is causing the disease; however, the bacteria proliferate so rapidly that generations can quickly create self-defense mechanisms to inhibit the effects of the antibiotic.

Participant A3: If he does not need it, then why is he going to have it? If something is not needed, it will atrophy.

Participant A11: The vermiform appendix used to serve a purpose, but as we no longer use it anymore, the appendix began decreasing . . . there will be people who are no longer born with the appendix, as is happening with the wisdom tooth.

Such examples of statements contain indicators that the teachers in training have teleological conceptions of evolution since they understand that phenotypic innovation can be induced by use and disuse or by the general needs of the organism. Specifically, in the case of Participant A10, there is a possibility that only the discourse is teleological, whereas the conception is not necessarily so. However, if the delivered speech is teleological, it means that this in-service teacher has not yet acquired the ability to make his or her discursive choices accurate for expression in accordance with the scientific model.

It is possible to conclude, through the analysis of the answers, that every conception based on Lamarckian premises has a teleological explanatory structure (use and disuse and law of acquired characters) but that explanations based on selective premises (referencing mutations and natural selection) can also have a teleological structure in regard to supposed origin-directed mutations and to strict selectionism (the conception that understands all living traits as structures that were selected in the past to perform their current function in a past time).

Throughout the development of the research, which made it possible to obtain continuous discourse of the teachers in formation, it was possible to verify that some of the teachers attribute a ubiquitous character to natural selection, which they treated as a process that occurs incessantly regardless of the circumstances. Some statements are presented below to highlight this statement:

Participant A5: Then, if you do not select it, why will you perpetuate it? I do not know, but if I did not make the selection . . . It is because he would not be selected and, hence, he would not pass on his genes.

Participant A3: It has to undergo a mutation, and this difference is what will be selected later.

Participant A5: So, if he is not selected, he will stay there with his mutation, then?

Such discourses highlight how teachers advocate the need to always include selective processes in any evolutionary history. It can be assumed that students did not understand that some requirements need to be met for natural selection to occur, such as the existence of variability among organisms in a given population. The viability of the occurrence of the natural selection process requires differences in the fitness of different organisms; however, as was indicated by the conceptions of the subjects that were surveyed, some students understand that if a trait is present in a living organism, it is because this trait was selected (which restricts selection).

In table 5.1, we aim to numerically illustrate the frequency of verbal teleological formulations in each of the stages. They are more frequent in the initial stages and decline in subsequent stages.

The following lines highlight the presence of the speeches that are classified in T1:

Participant A4: I think that nature has a great deal of pollen; it is better to have more left over than to be lacking [referring to the flower with unnecessary pollen because it performs parthenogenesis].

Participant A2: And, if the flower is not going to have an evolutionary tendency for the future, can it be understood that it will have pollen that is fertile, rather than sterile? [Referring to the flower with unnecessary pollen because it performs parthenogenesis].

Participant A1: And, if the flower needs to use these structures again later . . . I do not know [referring to the flower with unnecessary pollen because it performs parthenogenesis].

Participant A9: If she were dependent on having eyes, she would develop other senses to continue being able to see in that environment [referring to organisms living in caves].

Participant A10: At the moment, it seems not. We do not know how long this kind of mutation can last, or if it will occur more often. There is no way to predict if there will be demand for it in the distant future. For now, it does not seem good for the way we live. [Referring to the anomalies that occur in human embryos, the participant tried to defend a future utility to explain the occurrence of the mutation].

Participant A2: But in nature, it is obvious that she will prefer to have more females than males, because a male can copulate several females.

Participant A5: Because she never needed this fruit, he obtained another food source. [When asked about the lack of claws of a mammal that would allow them to climb fruit trees, the student argued that he did not develop the claws because he did not need it].

The following lines highlight the presence of speeches classified in T2:

Participant A4: Whenever you talk about evolution, improvement comes to mind. [The participant attempts to conceptualize evolution by such consideration].

Participant A1: I think it goes through all of the complexity from the simplest to the most complex. [The participant attempts to conceptualize evolution by such consideration].

CONCLUSIONS

In terms of the nature of pedagogical strategies, we can conclude that (a) the students must express themselves so that the teacher may obtain indicators of their conceptions (before, during and after the learning process) to carry out interventions; (b) in a monologue class model, in which only the teacher speaks and the students listen, the evolutionary mechanisms can be explained and exemplified, but if the teleology of the conceptions of the students is not problematized, the teleological obstacles may still remain; (c) it is necessary to challenge the traditional model of expository teaching, to allow the students to express themselves and resignify their conceptions through pedagogical interventions.

The relevance of this research lies in the selection of conceptual questions in the context of pedagogical strategies that facilitate the diagnosis of teleological obstacles and their resolution, such as the nature of genetic inheritance, the nature of the origin of genetic and phenotypic innovation, the distinction between the individual and population dimension, the distinction between the origin of variability and selection of variability, the distinction between physiological and evolutionary adaptation, the distinction between innate behavior and the inclusion of explanations for living traits besides natural selection (through discussions of genetic drift, pleiotropic genes, and evo-devo, for example).

The research focused on the study of teacher education, but the implications are not restricted to this target audience because it is an issue that may involve learning difficulties in various age groups. Thus, the strategies used here can be adapted and adopted by other teachers at different levels of education.

REFERENCES

Bardapurkar, Abhijeet. "Do students see the 'selection' in organic evolution? A critical review of the causal structure of student explanations." *Evolution: Education and Outreach 1*, no. 3 (2008): 299–305.

Bardin, Laurence. *Análise de conteúdo*. Translated by Luís Antero Reto e Augusto Pinheiro. São Paulo: Edições 70 LDA, 2011.

Brumby, Margaret. "Problems in learning the concept of natural selection." *Journal of Biological Education 13*, no. 2, (1979): 119–122.

Campbell, Neil A., Jane B. Reece, Lisa A. Urry, Michael L. Cain, Steven A. Wasserman, Peter V. Minorsky, and Robert B. Jackson. *Biology* (8th ed.). New York: Pearson, 2010.

Dawkins, Richard. *O Relojoeiro Cego*. Translated by Laura Teixeira Motta. São Paulo: Companhia das Letras, 2001.

Ferrari, Michel and Michelene T. Chi. "The nature of naive explanations of natural selection." *International Journal of Science Education 20*, no. 10 (1998): 1231–1256.

Futuyma, Douglas J. *Evolution* (2nd ed.). Sunderland, MA: Sinauer Associates Inc., 2009.

Gregory, T. Ryan. "Understanding natural selection: Essential concepts and common misconceptions." *Evolution: Education and Outreach 2*, no. 2 (2009): 156–175.

Gregory, T. Ryan, and Cameron A. Ellis. "Conceptions of evolution among science graduate students." *BioScience 59*, no. 9 (2009): 792–799.

Kelemen, Deborah. "Teleological minds: How natural intuitions about agency and purpose influence learning about evolution." In Rosengren, Karl S., Sarah K. Brem, E. Margaret Evans, and Gale M. Sinatra. *Evolution challenges: Integrating research and practice in teaching and learning about evolution*, 66–92. Oxford: Oxford University Press, 2012.

Lombrozo, Tania, and Susan Carey. "Functional explanation and the function of explanation." *Cognition 99*, no. 2 (2006): 167–204.

Mayr, Ernst. *What Makes Biology Unique?* Cambridge: Cambridge University Press, 2004.

Meyer, Diogo, and Charbel Niño El-Hani. "A evolução da teoria darwiniana." *ComCiência 107* (2009): 1–4.

Meyer, Diogo, and Charbel Niño El-Hani. *Evolução: O sentido da biologia*. São Paulo: Editora da UNESP, 2005.

Mossio, Matteo, and Leonardo Bich. "What makes biological organisation teleological?" *Synthese 194*, no. 4 (2017): 1089–1114.

Reece, Jane B., Lisa A. Urry, Michael L. Cain, Steven A. Wasserman, Peter V. Minorsky, Robert B. Jackson. *Campbell Biology* (10th ed.). New York: Pearson, 2015.

Sepulveda, Claudia, and Charbel N. El-Hani. "Obstáculos epistemológicos e sementes conceituais para a aprendizagem sobre adaptação: uma interpretação epistemológica e sociocultural dos desafios no ensino de evolução." *Acta Scientiae 16*, no. 2 (2014): 237–263.

Sepulveda, Claudia, Eduardo Fleury Mortimer, and Charbel N. El-Hani. "Construção de um perfil conceitual de adaptação: implicações metodológicas para o programa de pesquisa sobre perfis conceituais e o ensino de evolução." *Investigações em Ensino de Ciências 18*, no. 2 (2013): 439–479.

Silva, Paloma Rodrigues da, Mariana A. Bologna Soares de Andrade, and Ana Maria de Andrade Caldeira. "Biology Teachers' Conceptions of the Diversity of Life and the Historical Development of Evolutionary Concepts." *Journal of Biological Education 49*, no. 1 (2015): 3–21.

Toepfer, Georg. "Teleology and its constitutive role for biology as the science of organized systems in nature." *Studies in History and Philosophy of Science Part C: Studies in History and Philosophy of Biological and Biomedical Sciences 43*, no. 1 (2012): 113–119.

Vosniadou, Stella. "Conceptual change and education." *Human Development 50*, no. 1 (2007): 47–54.

Vosniadou, Stella. *International Handbook of Research on Conceptual Change*. New York: Routledge, 2008.

Yin, Robert K. "*Pesquisa qualitativa do início ao fim*." Translated by Daniel Bueno. Porto Alegre: Penso Editora, 2016.

Chapter Six

Darwin's Discussion of the Origin of Electric Fish

A Teaching and Learning Sequence in Youth and Adult Education

Gerda Maisa Jensen and
Maria Elice de Brzezinski Prestes

This chapter presents partial results of empirical science education research based in a Teaching Learning Sequence (TLS) (Méheut and Psillos 2004; Zabala 1998). The target public was middle school students of the Youth and Adult Education (EJA, in the Portuguese acronyms for *Educação de Jovens e Adultos*). The pedagogical intervention was an inquiry learning approach of an episode of the history of biology to teach the theory of biological evolution and its main mechanism, the natural selection. Only a general approach to evolution appears in the national as well as in the state of São Paulo curriculum guidelines for secondary education in Brazil, both for regular and special education. As Vilanova and Martins (2008) pointed out, the curricular proposal for Youth and Adult Education of the Ministry of Education (MEC, in the Portuguese acronyms) emphasize science content that focus on health, sexuality, and environmental problems. Biological evolution only appears among the discussion of general scientific ideas compared with mythical and religious ones and linked with the themes of the origin and evolution of the universe and life (Brasil 2002; Brasil 2006).

The city of São Paulo guidelines for EJA proposes more specifically the evolution of living beings, but, again, only through general and broad concepts of process and phenomena (São Paulo 2010). The São Paulo curriculum guidelines recommend a more detailed approach to the synthetic theory of evolution only in the high school level, and for the last year, after the study of genetics. High school in Brazil follows nine years of primary school, and is composed of three years. In each of them, there are annual courses in biology, as well as in chemistry and physics. Besides these official guidelines, a similar succession of the biological contents appears in textbooks, including those approved by the MEC National Book and Teaching Material Program

(PNLD, the acronym in portuguese for the *Programa Nacional do Livro Didático*) distributed to students of the state public schools across the country.

As argued by Nelio Bizzo and Charbel El-Hani, this sequence of content carries both a historical error and an epistemological error. Chronologically, theories of evolution are not posterior to genetics, and the knowledge about the causes of variations among species is not decisive nor even necessary for the proposition of the biological evolution and its main principle of natural selection. The authors suggest reducing the genetic contents in favor of a broader treatment of macroevolution, allowing the introduction of biological evolution to an earlier position in the curriculum (Bizzo and El-Hani 2009). Following this perspective, the pedagogical intervention developed for this research introduced the principle of natural selection in a much earlier stage of the formal scientific literacy generally practiced in Brazil, in the middle school level (EF 2, the Portuguese acronyms for *Ensino Fundamental 2*), corresponding to the fifth through eighth grades in the United States.

The motivation for the research (Jensen 2016), which generated the partial results explored in this chapter, came during a previous study of the first author to obtain the master's degree in history of science (Jensen 2008). The objective of that previous study was to trace the historical analysis about the nature of the phenomenon caused by some kinds of fishes in their preys, as well as anyone who touches them. This phenomenon was known from antiquity between fishers and residents near the sea or rivers and lakes. In the absence of explanation, many people thought this was a "fictitious" phenomenon. Natural philosophers of the seventeenth century were motivated to dedicate studies to the subject, developing alternative corpuscular and mechanical explanations. The "pain and numbness" caused by certain types of fish in their prey and in the hands of those who touched them, as the words used a century later by the Italian naturalist Lazzaro Spallanzani (1729–1799), led him to resume the investigation of the anatomy of these fishes. Spallanzani also performed experiments and discussed the case with contemporaries. At his time, the development of the studies on the nature of electricity led to the proposal of an electrical explanation for the phenomena manifested by these fish.

The previous historical research ended at this point, but a new question started to provoke the continuity of the studies. Spallanzani (Prestes 2003) like several other naturalists was gathering knowledge about the extinction of species, and some of them started to propose ideas about the modification of species; themes that, together with electricity, entered in the investigations of the nineteenth century. What would happen with the two centuries debate on the nature of the electric fish with the arrival into the scene of a new theoretical perspective, such as the theory of biological evolution proposed by Charles Darwin (1809–1882)? Did Darwin pay attention to this particular case among animals? The reading of Darwin's *The Origin of Species* quickly gave the clues for the answers, as will be detailed in the next section. At the same time, the intention was to explore this history of biology's episode as an interesting problem for science classes, along with the challenge of working the concept of natural selection with middle school students.

The motivation was the belief that is introducing the concept of natural selection at this stage, would allow students to develop and deepen their understanding of evolution throughout schooling years and favor the knowledge of the more systematic and later approach to the synthetic theory of evolution. Besides, since the electric fish case is only scarcely the subject of discussion, whether in the history of biology itself (Pauly 2007) or the textbooks of biology for any level of education; in Brazil, it could offer a genuine opportunity for students to think by themselves about the principle of natural selection. The study was carried on by one overall general and few particular hypotheses regarding the student's views, all of them searching for evidence of the relation between their religion or social-culture rooting beliefs about the proposed scientific problem (the nature and the origin of the fishes' electric organs). These hypotheses worked out as parameters to assess the learning process about the theory of evolution and natural selection by the end of the TLS.

THE HISTORICAL RESEARCH

The numbness and pain caused by some fishes was known and discussed since antiquity. The phenomena were undoubtedly known by the time, as has been already described by historians of science (Kellaway 1946). The torpedo was mentioned in one of the dialogs of Plato (427 BC–348 BC), stating that Meno "compares Socrates to a torpedo which paralyses everything which comes near to it" (Plato 1887, 19). Aristotle (384 BC–322 BC) made extensive anatomical observations on rays and eels, including comments on their habits such as that they bury themselves in the sand to catch creatures that swim in its way by narcotizing them "by the power that is resident in its body" (Aristotle 1887, Book IX). Aristotle adds that this phenomenon "was actually observed in operation," a kind of phrase he used to indicate that he relied on its truthfulness. There were also records in ancient times of therapeutic application of these fish. The roman physician Scribonius Largus (c. 1 AD–c. 50 AD) recommended placing a living black stingray in the region affected by headache (Baldwin 2008, 80–81). In the seventeenth century, however, the phenomenon presented by these fishes was considered by many to be fictitious. Thus, eels, catfishes, and torpedoes called the attention of the natural philosophers.

The Italian physician and naturalist Francesco Redi (1626–1697), "to be sure of the fact and to be able to speak with certainty of science" (Redi 1671, 17), asked fishermen to bring him a freshly caught fish. When he touched the fish, immediately felt a tingling in his hand, arm, and shoulder, with such an annoying tremor and a sharp pain at the tip of his elbow, causing him to withdraw his hand immediately. Following the detailed anatomical observation of the animal, Redi concluded that the "painful virtue" seemed to be caused by two sickle-shaped muscles which occupy the entire interior space of the body (Redi 1671, 17). His disciple, Stefano Lorenzini (1652–?), resuming the anatomical observations of the torpedoes, concluded that the numbness or pain "comes from many corpuscles, which come out of the torpedo and enter into the hands of those who touch it" (Lorenzini 1678, 113). In sum, among others, these authors attributed a corpuscular explanation for the phenomena.

Another Italian naturalist, Giovanni Alfonso Borelli (1608–1679), criticized the idea of the emission of corpuscles as merely imaginary and described that when touched, the fish stirs and causes a violent blow, which creates a painful numbness on the hand that reaches it. In other words, the unpleasant effect was due to the blast, meaning that its explanation was mechanical (Jensen 2008). The French naturalist René-Antoine Ferchault de Réaumur (1683–1757) also rejected the corpuscular interpretation and was more aligned with the mechanical cause. However, instead of the violent blow proposed by Borelli, Réaumur considered that the effect would be due to an imperceptible movement, so fast, that the sickle-shaped muscles suffered, alternating from a concave to a convex position (Réaumur 1714).

Réaumur's explanation prevailed until the phenomenon began to be analyzed from the perspective of the electrical studies developed in the late eighteenth century and nineteenth century. By that time, with the development of local natural histories, naturalists such as the French Michel Adanson (1727–1806) in his voyage to Senegal described a new species of a fish known by the natives of that country as "ouaniear" that he called "trembleur" (trembling). He described the effect of the very painful quivering felt by those that touched the fish as being similar to that caused by the new "electric" instrument, the Leyden jar (Adanson 1757, 135). The same occurred to English physician Edward Bancroft (1745–1821) referring to a torporific eel he found on the river Essequebo in Suriname, South America. For Bancroft, the transmission of the torporific or electrical particles occurred through the water where the eel swam (Bancroft 1766). An experimental study on the phenomenon caused by the torpedo was carried on by John Walsh (1725–

1795) and published in the *Philosophical Transactions* in 1773. The work is composed of two letters addressed to the American researcher and diplomat Benjamin Franklin (1706–1790). Walsh did not even mention the hypothesis of the mechanical shock of Réaumur. Instead, from the first paragraph, Walsh explained the phenomenon by the electricity framework (Walsh 1773).

Meanwhile, in the passage from the eighteenth to the nineteenth century, the electric fish also appeared in another conceptual context discussed by the naturalists, the modification of species. In the second volume of the *Histoire Naturelle des Poissons*, Comte de Lacépède (1756–1825) mentioned the electric fish and other cartilaginous fishes. Lacépède states that these fishes announce the uniqueness of their origin, testifying "the unity of the model according to which nature has shaped all the species of fish." Nature, he continues, "created only one living being, from which she multiplied copies more or less modified" (La Ville-sur-Illon, Bernard-Germain-Étienne, and Conde de Lacépède 1798, 129).

In 1802, Etiénne Geoffroy Sain-Hilaire (1779–1844) made comparative anatomical studies of the electric organs in torpedo ray, eel, and catfish from the Nile (Saint-Hilaire 1802). Considering "nature confined within certain limits," so that "a single plane forms all living things." Saint-Hilaire supposed the similarity between the electrical organs as a possible reason to assume that these fishes belong to neighboring groups. However, they were very different genres (Saint-Hilaire 1796). Electric organs occur in fishes belonging to different groups not related to each other; when Charles Darwin wrote *The Origin of Species* he mentioned these organisms among the special difficulties that remained not explained by his theory of evolution. After all, how can we explain the origin of these organs through natural selection from a common ancestor if they are not close to each other?

Invocating the far too ignorance about the subject, Darwin had to be content with an analogy. Just as two men can independently arrive at the same invention, natural selection would also similarly modify organs of beings unrelated hereditarily (Darwin 1859, 193–194). So, this was the difficulty faced by Charles Darwin's evolutionary interpretation of the phenomena: since electrical fishes were very different from each other, meaning that they did not have a common ancestor, these organs should be new. Moreover, there was no evidence in the geological record that the electrical organs would have disappeared in other types of fish "through disuse or natural selection" (Darwin 1859, 193). Darwin continued to think about this difficulty, and in the fourth edition of the book, in 1866, he added some arguments and concluded that different common muscles had undergone modifications over time, transforming themselves into analogs electrical organs. He also added more evidence to the idea of natural selection as the cause of organ transformation from the Fritz Müller (1821–1897) book, *Für Darwin*, published in 1864.

Müller studied the out-of-water breathing apparatus of two species of crustaceans that belonged to two different families. These air-breathing apparatus were very different from each other. For Müller, the apparatus would have been modified over time if different species had gradually adapted to life and breathing out of the water. "By what intelligible breeding act would have arisen two different air breathing apparatus if most of the anatomy of crustaceans is very similar to that of other crustaceans?" (Müller [1864] 2009, 59). In the sixth edition of the *Origin*, Darwin added that organs of distant beings must develop independently for the same function and it happened due to two simultaneous factors, the nature of the organism, in this case, the common muscles in different species of fishes, and the surrounding conditions (Darwin 1876, 153).

Darwin admitted that the other (minor) difficulties posed by the electric fish still prevented him from knowing by what gradual steps the muscular organs of certain fishes would have turned into electric organs. First, the absence of transitional forms in the known species of electric fishes; second, the absence of geological records that could show the existence of other kinds of fishes with electric organs that disappeared, either by *disuse* or by natural selection. In a letter to George Henry Lewes (1817–1878), however, he explicitly mentioned that the electrical organs of the fish could have originated from an electrically weaker discharge and related to other functions in the ancestral fish (Darwin 1868, 4). Thus, despite the lack of more in-depth knowledge on the subject of his time, Darwin was sure that natural selection could, with different materials and variations, achieve the same functional result, so that, in the end, the fish's electrical organs did not represent an objection to his theory anymore.

The first minor difficulty that Darwin raised received an answer a century later. In 1958, a fish of turbid water with a weak (meaning intermediate) electric discharge used to localize its prey was described (Lissmann 1958). Nowadays, many similar cases are well known. Recently, molecular research made a new theoretical and methodological approach to electric fish (Gallant, Hopkins, and Deitcher 2012). The presence of similar characteristics independently acquired through biological evolution, not representing genetic proximity between different species, is known by the concept of "homoplasy" (Ridley 2013, 454). As mentioned earlier, this historical episode guided the planning of a TLS for a particular context, the EJA modality of education, more detailed in the next section.

EMPIRICAL RESEARCH: EJA MODALITY OF EDUCATION

The Youth and Adult Education, EJA, is an official modality of education in Brazil devoted for those who did not have access or continuity of studies in elementary and secondary education in the proper age (Brasil 1996). Students must be fifteen years old to begin EJA for the elementary and secondary school level, and eighteen years old to start EJA for the high school level. It is important to note that besides adults that never attended or left school long ago, EJA includes teenagers and young adults excluded from the system due to school repetition and premature entry into the labor market. These two distinct student profiles, young and mature adults, are characteristic of EJA classes. In general, these students are workers, although many of them unemployed due to the enormous economic crisis of recent years, with the mark of 12.7 million unemployed in January 2019 (IBGE 2019).

Many kinds of diversity characterize the EJA students in the city of São Paulo (São Paulo 2004, 7). They have different ages, from fifteen to sixty or seventy years old, even though about 43 percent of this public are between fifteen and twenty-five years old, are single but have family responsibilities. They have different life experiences, coming from rural or urban areas. They have a diversity of cultural baggage brought from the particular region of the country they came—remembering that São Paulo, the most developed and populated city of the country, is an attractive pole of internal migration, mainly, but not only, from the Northeast region of the country.

On the other hand, students share some past experiences and present similar expectations. In general, these people do not have access to the literate culture (Possani and Silva 2004, 7). Many of them could not study in the normal age because they had to enter the labor market earlier. They have, especially older adults, a high expectation that schooling will enable their personal development, not just financially. The majority are Afro descendent. Regarding religion, 62 percent of the students declare themselves Catholic, and 24 percent, Evangelicals; 6 percent indicate other religions, and 8 percent say they do not practice any religion. The proportion of those who declare no religion is higher among adolescents (18 percent) (São Paulo 2004).

The EJA course lasts for four years to complete the Brazilian mandatory primary and secondary school curriculum (of nine years in the regular school). There are four stages in the course, grouped in two semesters. (1) Literacy Stage and Basic Stage, focusing the abilities of reading and writing in the Brazilian-Portuguese language and mathematics, with a generalist teacher. (2) Complementary Stage and Final Stage, with the first classes with specialist teachers of Portuguese language and mathematics, as well as of geography, history, arts, English, and sciences. Each stage, or one year course, has two hundred school days, and classes from Monday to Friday. Although basic education up to age seventeen is mandatory in Brazil, it is not yet universalized. According to data from 2018, 97.7 percent of the population aged six to fourteen years is enrolled in elementary education and only 67.5

percent of the population aged fifteen to seventeen is enrolled in secondary education (Anuário 2018).

EJA STUDENTS PARTICIPATING IN THIS RESEARCH

Usually, religiosity is perceived by this public (especially the older) as personal and confidential information. During the implementation of the TLS of the present research, however, the eleven students that attended the class that day ended up to reveal their religions: two students identified themselves as practicing Roman Apostolic Catholics; five as practicing Evangelicals; two atheists; two said: "I do not know." Evangelical students made a point of identifying the names of the churches to which they belong: Christian Congregation in Brazil (*Congregação Cristã do Brasil*); Church of the Power of God (*Igreja do Poder de Deus*); Church Vine (*Igreja Videira*); Universal Critical Church (*Igreja Universal Crítica*); Peace and Fullness Church (*Igreja da Paz e Plenitude*).

THE TEACHING LEARNING SEQUENCE (TLS)

A TLS is a series of teaching and learning "activities ordered, structured and articulated for the achievement of certain educational objectives, which has a beginning and an end known to both teachers and students" (Zabala 1998, 18). The historical episode described in the previous section guided the general structure of the TLS planned as a close interaction between interventionist research and an empirically tailored product for students, carefully integrating the teaching practice and the targeted learning (Méheut and Psillos, 2004). The TLS was elaborated by the teacher-researcher (the first author of this chapter) for her students. It followed the theoretical and methodological orientation of the curricular system of the school, focusing, historically and socially, its specific culture.

The teacher-researcher had a long teaching career and experience with EJA, feeling high motivation to develop and implement an innovative didactic proposal. The planning of the TLS was based in this previous experience and complemented by a systematic collection of data on the profile of students, human resources, and materials available at school, curricular guidelines and objectives of the municipal EJA education network, the political-pedagogical project of the school, ethical aspects of research in education. Besides this practical and contextual information, a review of the literature on empirical research in science education informed the methodological treatment.

DIDACTIC STRATEGIES AND INSTRUCTIONAL MATERIAL

An inquiry approach to the historical episode guided the elaboration of specific instructional materials. Small texts, written by the researcher, of different moments and aspects of the historical study, short excerpts from primary sources, and written activities posing questions in the manner of the inquiry teaching and learning process. It is important to remember that the target audience was poorly literate, and all instructional material needed to be drawn up in function of this condition (concise texts, lots of drawings, clear and direct messages). Through the inquiry approach, scientific contents appeared as questions related to the cause and origin of the electric fish organs, as well as some concepts of the theory of biological evolution. In the structure of the TLS presented bellow these Inquiry Questions (IQ) received numbers according to each meeting (or class).

The format of the IQ followed Allchin (2013) and Berçot and Prestes (2016). They are direct question inviting students to express their own reflections and to situate them as the protagonists of the action ("what you would do or plan or think," instead of "what should be done"). They inquiry about specific issues, linked to particular events ("If you were in the place and time of those scholars of the seventeenth century, what would you do to decide if those strange fishes were fictitious or real?" or "If you were in the position of Darwin, what would you suggest to explain the origin of the electric organ in different species of fishes?" instead of "what is the causes of pleiotropy?").

The teacher-researcher used oral communication to pose the questions, images, and written study scripts. She invited students to discuss and express their suggestions to solve each presented problem through oral discussion in small groups and to fill the scripts with particular written questions and drawings. Only after the exploration of the students' views for the first question, the teacher-researcher presented the different answers from the history of biology and then followed to the next problem—a methodological teaching approach that balances the open questions that characterize the inquiry learning with the path effectively traversed by the scholars of the past (Allchin, 2013). The formulation of the IQs was followed by the formulation of probes designed for the joint interview, as it will be detailed below. The probe developed by Southerland, Smith, and Cummins (2005), as a previously planned question for an interview or an instruction in a written activity seeks to make students explicitly explain, with their own words, what they understood about a given concept.

OBJECTIVES OF THE TLS

Concerning the learning of science, the primary objective of the TLS was to promote the learning of the two main aspects of the theory of the biological evolution, the idea that species descend from common ancestors and that they undergo modifications that are mainly, but not only, a consequence of natural selection. Darwin's observations on the variations (domestic and under nature) and its causes were a second objective.

Concerning the learning of the nature of science, the primary objective was to enable students to recognize the existence of different scientific explanations for a given scientific issue over time, resulting in changes in accepted scientific knowledge. The focus was to discuss some of the components taken into account to build scientific explanations, such as empirical evidence and theoretical context. Concerning the learning of the history of science, the primary objective was to get in touch with an episode of the history of biology, involving different scholars over a long period (from the seventeeth to nineteenth century), relating to their epistemic and socio-cultural context. Concerning the overall learning process in EJA, all the work was also intended to develop the reading and writing competences of the students.

STRUCTURE OF THE TLS

The TLS is composed of four meetings, each one lasting about ninety minutes. Each meeting is described below.

First Meeting

The theme for the first meeting is the electric fish and its studies in the history of biology. The contents, strategies, and inquiry questions (IQ) are as follows: The teacher-researcher presents a review from the first block of classes and an overview about current well-known electric fishes, including the Brazilian species, showing their effects when touched by other organisms without reference to the causes of the phenomenon. The instructional material used is primary sources of images. The inquiry questions are: IQ 1.1: If you know someone that does not believe that there is such a kind of fish (thinking they are fictitious), what would you propose to convince him or her about their existence? Group-class discussion. IQ 1.2: What kind of explanation would you give for what was the cause of the phenomenon? A grouped class discussion of the students' ideas is conducted. The teacher-researcher presents the strategy used by the seventeenth-century natural philosopher Francesco

Redi to recognize the very existence of these fishes. She also explains the anatomical studies made by Redi and other scholars of the following century identifying the origin of the effect in the sickle-shaped muscles of the fish and formulating different theoretical models (corpuscular, mechanical, and electrical) to explain it. Students are invited to indicate similarities and differences between their models and that natural philosopher's models. Students are invited to propose their explanations for the phenomenon manifested by the electric fish and compare the different models offered by the seventeenth- and eighteenth-century natural philosophers with their own. The hypothesis for this meeting is that students recognize that the phenomenon is electrical but do not go much further indicating internal organs or models of explanation. Some students could attribute the event to God's wish.

Second Meeting

The themes for the second meeting are new century, new facts, new models, new knowledge on living beings. The contents, strategies, and inquiry questions are as follows: The teacher-researcher distributes an instructional material with fishery schemes showing great variety among individuals of the same population and species. The inquiry questions are: IQ 2.1: Let's consider a hypothetic fish population that have adults with different sizes, some are small and some are large. When fished, the larger ones, that have sales value, are sent to the market, and the smaller ones are returned alive to the lake. The fishes returned to the lake will reproduce. How do you think will be the size of the descendants? Students' discuss in small groups and present their opinions to group-class. IQ 2.2: If the fishermen always return the smaller fish to the lake, what happens to the average size of the fish over the years? For how long? Who is responsible for the fish size selection? After students' discussion, at the end of the class, the teacher-researcher calls their attention that this notion is named, in the present day theory of biological evolution, "artificial selection." With the objective being to understand the notion of artificial selection, it is hypothesized students should say that fishermen are responsible for the smaller size of the fish population over time in the region, naming it or not as "artificial selection." Some students could also say that, no matter what the fishermen do, the fish population is a result of God's wish.

Third Meeting

The theme for the third meeting is the comparison between artificial and natural selection. The contents, strategies, and inquiry questions are as follows: After a review to help students that missed the previous classes, the teacher-researcher distributes a short excerpt of chapter 1 of *The Origin of Species* (in Portuguese) about domestic pigeon varieties together with drawings of them that illustrate the idea of Darwin about artificial selection. The drawings of pigeons were copied from an article by Roberto Martins, who also details the use of the example of pigeons in Darwin's argument. Among other domestic animals and plants, Darwin discussed the domestic pigeons, which were highly prized by English creators of the nineteenth century. "If the man had been able to transform the *Columba livia* into races as different from pigeons," such as the images shown to students, then, Darwin argued, "nature is also capable of producing much larger modifications over the millions of years available. [. . .]" Thus, the common origin of all breeds of domestic pigeons is a fundamental part of the argument put forward by Darwin in *The Origin of Species*. Darwin studied various domestic animals and plants, especially the domestic pigeons, which were highly prized by English creators of the nineteenth century. To develop this part of his contribution, Darwin devoted a few years to the creation and study of domestic pigeons, performing analyzes of their skeletons and also cross-breeding experiments ending by evidencing "the power of the natural selection for the creation of new species and genera" (Martins 2012, 115). The guiding inquiry questions are: IQ 3.1: What do you think that led many breeders to accept that all variations of domestic pigeons descended from the Rock Pigeon (*Columbia livia*, Gmelin 1789)? IQ 3.2: What do you think about the existence of species'variations that are not domesticated and live in nature, such as the rufous-bellied thrush (*sabiá*, in Portuguese, *Turdus sp*)? Students are invited again to read the excerpt of Darwin's book, this time to underline the passages referring to the concepts discussed in this meeting: artificial selection, natural selection, random variation, favorable variations. Next, students should say or write what they think about the existence of domestic and natural varieties. The objective of this meeting is to understand Darwin's analogy between artificial and natural selection. Students could say that the breeders are responsible for the variation of the domestic pigeon, while the variation in nature is the result of God's Word described in Genesis or they could say that it is the result of the working of nature by itself, since God provided it such power. Maybe some students do not refer to God and answer with the natural selection explanation.

Fourth Meeting

The theme of the fourth meeting is the origin of the electric organs in different kinds of fish. The contents, strategies, and inquiry questions are as follows: The teacher-researcher gives instructional material with primary sources with illustrations of the fishes. She proposes the analysis of the possible origin of the electric fish organs as an example of Darwin's notion of the relation between variations and struggles for survival. The guiding inquiry questions are: IQ 4.1: How would you explain the origin of the electric fish, torpedo? IQ 4.2: How could you justify the existence of many different kinds of electric fishes? Could these different kinds of fish have the same ancestor? The teacher-researcher using Saint-Hilaire's figure explains the differences between the body of the different species of fishes and the different locations of the electric organs in them. The students should answer that the origin of the electric organ and the variety of species are due to natural selection, but some of them could keep the God's cause argument or could create a conceptual profile combining scientific and religious reasons.

VALIDATION OF THE TLS

The validation of the TLS focused on student's cognitive pathways all through the learning process and was previous to its implementation in the classroom, as proposed by the framework of the Didactic Engineering (Méheut and Psillos 2004). The *a priori* validation followed exploratory questions with three groups of people: a) the members of the Research Group on History of Biology and Education of the Institute of Biosciences (IB) of the University of São Paulo (USP), in the first semester of 2014; b) some teachers of the school (including teachers of other disciplines, members of the school council, the principal, and the pedagogical coordinator), in the second semester of 2014; and c) the ethics committee of the IB-USP. The school council is a collegiate with a deliberative function with the mission to defend the interests of the purposes and objectives of public education in the city of São Paulo (São Paulo Municipal Law 14.660/2007). Answers to the exploratory questions and notes taken by the teacher-researcher were analyzed and incorporated in the TLS whenever they appeared to contribute to the improvement of TLS objectives and approaches. Particularly relevant were the specific, yet rather punctual, contributions on instructional materials.

METHODOLOGY

Implementation of the TLS

The TLS was implemented in May 2015 in the discipline of sciences for students of the first semester of the complementary stage of EJA, corresponding to the sixth year of the regular Brazilian modality of school, or the fifth grade in the United States. They followed night classes, from 7:00 p.m. to 11:00 p.m. It is important to note that some of the students never attended a school before, so they were in their very first contact with formal science classes. In other words, they were acquiring their scientific literacy. The space used was the regular school classroom.

Data Collection

The data came from the written and pictorial production of the students, teacher-researcher's meetings notes, and the audio record of all four meetings, including what we may call a "collective interview" at the end of the last session. After each class meeting, the teacher-researcher analyzed the collected data, and coded each student's ideas about evolution. They were compared with previous indications recorded in the field notes and checked to see if they show discrepancies or not. If these discrepancies or any new sign pointed to conflicting conceptions, a semi-structured interview with those students would follow the last meeting, as demanded by the qualitative methodology. Based on the experience with EJA students, however, the teacher-researcher feared that students would be intimidated if they were alone to talk to her. The decision, then, was to conduct a kind of whole class review, pre-planned with specific probes directed for those eventual target students. Direct probes to these students and their manifested conflicts would be the focus of the collective interview.

There were fifteen students attending the course, but an average of ten students were kept at each meeting, and not always the same ones—this is part of the "culture" of EJA that, despite being an onsite modality of education, the attendance is mandatory in only 50 percent of the classes of a specific discipline. Data collected in every meeting, regardless of the number or which students were present, were analyzed. Whenever possible, students worked in small groups (three to four students), and each meeting ended with a final discussion involving the whole class.

Qualitative Analysis

The qualitative analysis of data consisted in the comparison between the initial hypothesis expressed in the TLS and the answers students presented (written and oral) to different IQs as well as with the probes used in the group-class interview at the end of the TLS. The qualitative analysis focused on gathering some evidence about the fruitfulness of using a TLS based in a historical episode and through an inquiry approach to teaching the theory of biological evolution. The overall hypothesis was that these tools, history of biology and inquiry learning, could function as attractive elements for students, and could prevent them from resisting the scientific explanation for the origin of species, without confrontation with their religious beliefs.

In the first meeting, students recognized the existence of different kinds of electric fishes such as the electric eel, well known by them. Together, in the group-class, they managed to answer IQ 1.1, completing the phrase, "If I were a scholar of the seventeenth century" with their ideas:

> I would plan a trip to a beach town to observe this fish named torpedo considering the length of the trip at the time, the possible contingencies, the necessary financial resources. Being there, I would need to wait for the arrival of the fishermen from the sea, and this could take days. I could repeat myself the procedures of the fishermen traveling on a deep sea fishing boat and undertaking studies on the spot because of the difficulty of transporting aquariums at the time; many fishes die after fishing. I would need to plan how to collect, transport and keep electric fishes alive. Because if the phenomenon was electric, it might not happen after the fish was dead; all these ideas are correct, but at that time, I would take somebody with me as a witness. After, as a scholar, I needed to write an account to document all that I have done and published it.

The teacher-researcher had the perception that the adults' previous knowledge and life experiences in the Northeast Coast of Brazil contributed to this particular class discussion more richly than those occurring teenagers of the regular schooling.

In the second meeting, students reached a consensus regarding IQ 2.1 and IQ 2.2, stating that descendants will have different sizes, some will be small, and others will be large, and that the average fish size of this population would undoubtedly be lower. Here again, it seems that students based their responses on their own life experience. When asked how long it would take, adults presented a better sense of time when compared to the younger ones in the classroom. When asked who would be responsible for the selection of the fish size, all of them answered that it was the fishermen, the human being. Until the second meeting, contradicting the hypothesis for the specific sessions, no student referred to God's wish.

Regarding the IQ 3.1 of the third meeting, seven students seemed to understand it well enough and pointed to the human being as responsible for the variations of the present species. But at this point, three students found it difficult to understand the Rock Pigeon as the common ancestor of the domestic pigeon varieties, and the God creation argument appeared among them. To further explore their views, the collective interview at the end of the forth meeting was necessary, as detailed below.

Figure 6.1. Student torpedo's drawing. The legend lists, from the top to the botton: head, electric muscle, tail, sting and fin.
Oral consent was obtained to reproduce this drawing done by a Portuguese student.

Regarding the fourth meeting and IQ 4.1, most of the students recognized that the fish's phenomenon is muscular and electric in torpedoes. They showed no resistance to explain the origin of such organs through natural selection from common ancestry. One example of this understanding is the drawing made by one of them in figure 6.1.

As mentioned before, at the end of the fourth meeting, oral statements, written texts, and drawings from the previous ones were added by answers of three particular students to probes planned for a "collective interview." The necessity of extra data emerged at the third meeting when three students manifested their creationist conceptions and displayed little understanding of Darwin's ideas.

The triangulation of the audio records of the joint interview with the previous data showed that the difficulties of these three students (students 1, 2, and 4 in the following transcript) seem to come from different factors. One of the students seemed to have had a little more difficulty remembering the terms, as highlighted in italics in the quote of student 1 below.

> *Teacher-researcher:* After showing for a particular student his writing material, where he wrote "Natural selection with rock pigeon, with species variation," she asked for this very student: Why did you call it "natural selection"?
>
> *Student 1:* Well, because a part of nature did it.
>
> *Teacher-researcher:* But why did these varieties arise? Was there interference from the human being?
>
> *Student 1:* Yes.
>
> *Teacher-researcher:* When men make the crosses and choose an exciting characteristic, what is it called?
>
> *Student 1: Yes, I am trying to remember. Because in the trees it would be like the grafting of the plants. Now in this one, I can't remember.*

The other two students, however, presented a different position. These students were not present in the class when they spontaneously touched the subject and the teacher-researcher was able to make the records of their religions from their own statements. However, from indirect clues, the teacher-researcher is quite sure that student 2 is a practicing Roman Catholic. Student 4 sustained an Evangelical discourse, but he did not identify himself as such. Student 2, insisted on resorting to the biblical explanation, as his highlighted answers below show. Student 4, by his turn, expressed oscillatory interpretations, in the manner, Eduardo Mortimer called a "conceptual profile" on the subject (Mortimer 1995). His highlighted answers below indicate it well.

> *Teacher-researcher:* How could you explain the origin of the electric fish, torpedo? And the others?
>
> *Student 2:* [. . .] *the animal was criated. Criated by the Lord.*
>
> *Teacher-researcher:* Let me ask the whole class, God had created these fishes too?
>
> *Student 2: He did.* That is the same to ask us: who came first, the egg or the chicken.

Teacher-researcher: You have said earlier that God created all kinds of plants and animals, that they already existed in the time of Noah. And that when God sent the flood, Noah placed a couple of each type inside the ark. Noah also included a couple of fish?

Student 4: No, the fish didn't need to go inside the ark. They live in water.

Teacher-researcher: So, this means that nowadays fishes are descendants of those first fish created by God before the flood? And that they are identical to that first fish?

Student 4: No. Because the fish was then created, but the present generations of fish are a mixture of races, and there maybe it got to be electric at some point—sometimes there were no electric fish. But we know that they [scientists] are cloning. They take a type of animal, take the semen there and can put it on any other animal that is not of the breed, and a son can be born [. . .] *of a race. And the fish was born. Sometimes the fish breeds with another and there, sometimes, the type of fish changes. If it were a catfish, it would not come out so catfish, with another species it already comes out a "different type" of catfish, a type that gives a shock. So the fish can not be the same. Then they evolved with each other.*

Teacher- researcher: What proof or evidence we could look for to say that there are fishes with electric organs that disappeared? Because, if you are a scientist, you must have to look for evidence to convince the others, right? What would you look for?

Other students in the class: Fossils, eggs, mortal remains.

Student 4: No, teacher. God created the electric fish, but he did not create just one kind of electric fish. The eel is the eel; the stingray is the stingray [. . .]

Teacher-researcher: You had said to me that Noah placed a couple of each animal and plant species in the ark. Where, in the Bible, is written that there was more than one kind of fish?

Student 4: No, teacher, in the Bible it speaks like this: God created the fish and gave name: Fish; God created the Birds and gave name: Birds; God created animals.

CONCLUSIONS

The theory of evolution is not well known nor well understood by the general public, like the youths and adults at the beginning of the research classes described here. What would be the reaction of EJA students to the teaching of the biological evolution theory? Since they are adults, and with different religious backgrounds and commitments, would they feel comfortable to talk about this subject? If they accept to discuss it, would they confront the scientific explanation with their own beliefs?

General questions like that motivated this research. The literature on students' views about evolution and creationism is vast. However, little is known yet about the opinions of students in the youth and adult special education. This research aimed to find some answers as exploratory scenario added by an innovative approach to this target audience, a TLS based in a historical episode treated as an inquiry-based learning approach. If some positive evidence appeared, the research could encourage new investigations and curriculum improvements for the EJA system.

Seven students showed increasingly more scientific answers to the IQs and the probes along the four meetings in a manner that they reached the main science learning objectives of TLS; namely, understanding the concept of common descent and the principle of natural selection, as well as the analogy between natural and artificial selection. So, the history of biology episode coupled with inquiry learning strategy worked well and could be a key to achieve scientific literacy, at least with the same kind of public.

In parallel, as the excerpts of the third and fourth meetings show, two students did not achieve the same learning performance. One remained resistant (student 2) while the other (student 4) attempted to construct a conceptual profile to accommodate the scientific explanation with the religious conception. These two students did not perform the desirable conceptual change from the religious understanding to the scientific explanation. But the good news is that the TLS has created an inviting environment for them to express their positions and at least partially attempt to incorporate scientific explanations for the phenomena they observe.

REFERENCES

Adanson, Michel. 1757. *Histoire Naturelle du Sénégal: coquillages: avec la relation abrégée d'un voyage fait en ce pays, pendant les années 1749, 50, 51, 52 & 53.* Paris: Chez Claude-Jean-Baptiste Bauche.

Allchin, Douglas. 2013. "Problem and case-based learning in science: An introduction to distinctions, values, and outcomes." *Life Sciences Education* 12: 364–362.

Anuário Brasileiro da Educação Básica. 2018. *PNAD Contínua*. São Paulo: Moderna.

Aristotle. 1887. *Aristotle's history of animals in ten books*. Trans. Richard Cresswell. London: George Bell.

Baldwin, Barry. 1992. "The career and work of Scribonius Largus." *Rheinisches Museum Für Philologie* 135, no. 1: 74–82.

Bancroft, Edward. 1759. "Letter II, Rio Demerary: An essay on the natural history of Guiana, in South America . . ." In: *Several letters*: 191–200. Accessed February 20, 2019. http://archive.org/details/essayonthenaturalhi00banc/page/nb.

Berçot, Filipe Faria, Maria Elice B. Prestes. 2016. "Narrativa de história da biologia para a sala de aula: Abraham Trembley (1710–1784) e a criatura que desafiou a classificação." *Boletim de História e Filosofia da Biologia* 10, no. 4 (December): 7–22. Accessed March 10, 2019. http://www.abfhib.org/Boletim/Boletim-HFB-10-n4-Dez-2016.pdf.

Bizzo, Nelio, Charbel El-Hani. 2009. "O arranjo curricular do ensino de evolução e as relações entre os trabalhos de Charles Darwin e Gregor Mendel." *Filosofia e História da Biologia* 4: 235–257.

Brasil. Ministério da Educação. Secretaria de Educação Fundamental. 2002. *Proposta Curricular para a educação de jovens e adultos: segundo segmento do ensino fundamental: 5ª a 8ª* série. Brasília: MEC/SEF.

Brasil. Ministério da Educação e Cultura. Secretaria da Educação Básica. 2006. *Orientações Curriculares de Ensino Médio. Ciências da Natureza, Matemática e suas Tecnologias: Orientações Educacionais Complementares aos Parâmetros Curriculares Nacionais-PCN+*. Brasília: MEC/SEB.

Darwin, Charles Robert. 1859. 1st ed. "Difficulties on theory." In *On the Origin of Species by Means of Natural Selection, or the Preservation of Favoured Races in the Struggle for Life*, 171–206. London: John Murray. Accessed February 20, 2019. http://darwin-online.org.uk.

———. 1866. 4th ed. "Difficulties on theory." In *On the Origin of Species by Means of Natural Selection, or the Preservation of Favoured Races in the Struggle for Life*, 189–226. London: John Murray. Accessed February 20, 2019. http://darwin-online.org.uk.

———. 1868. Charles Darwin to G. H. Lewes. Letter 6308. Darwin Correspondence Database. Accessed February 20, 2019. http://www.darwinproject.ac.uk/entry-4881.

———. 1876. 6th ed. "Difficulties on theory." In *On the Origin of Species by Means of Natural Selection, or the Preservation of Favoured Races in the Struggle for Life*, 133–167. London: John Murray. Accessed February 20, 2019. http://darwin-online.org.uk.

Gmelin, Johann Freidrich. 1789. *Caroli a Linné, systema naturae* (13th ed.). Tom. I. Pars II.—pp. 501–1032. Lipsiae.

Gallant, Jason R., Carl D. Hopkins, David L. Deitcher. 2012. "Differential expression of genes and proteins between electric organ and skeletal muscle in the mormyrid electric fish Brienomyrus brachyistius." *Journal of Experimental Biology* 215, no. 14: 2479–2494.

IBGE [Brazilian Institute of Geography and Statistics]. 2019. *Taxa de Desemprego* [Unemployment Rates]. Retrieved from https://ww2.ibge.gov.br.

Jensen, Gerda Maisa. 2008. "Lazzaro spallanzani (1729–1799) e o torpedo: um tipo de peixe elétrico?" Master thesis, Pontifical Catholic University of São Paulo.

Jensen, Gerda Maisa. 2016. "Charles Darwin (1809–1882) na Educação de Jovens e Adultos: história e natureza da ciência no ensino de ciências." PhD diss., University of São Paulo.

Kellaway, Peter. 1946. "The part played by electric fish in the early history of bioelectricity and electrotherapy." *Bulletin of the History of Medicine* 20, no. 2: 112–137.

La Ville-sur-Illon, Bernard-Germain-Étienne, Conde de Lacèpede. 1798–1803. *Histoire naturelle des poissons*. Paris: Chez Plassan, Imprimer Librarie. Accessed February 20, 2019. https://archive.org/search.php?query=histoire+naturelle+des+poissons&.

Lissman, Hans Werner. 1958. "On the function and evolution of electric organs in fish." *Journal of Experimental Biology* 35, no. 1: 156–191.

Lorenzini, Stefano. 1678. *Osservazioni intorno alle torpedini*. Firenze: L'Onofri.

Martins, Roberto de Andrade. 2012. "A origem dos pombos domésticos na estratégia argumentativa de Charles Darwin." *Filosofia e História da Biologia* 7, no. 1: 91–116. Accessed March 14, 2019. http://www.abfhib.org/FHB/FHB-v07-n1.html.

Méheut, Martine, Dimitris Psillos. 2004. "Teaching-learning sequences: Aims and tools for science education research." *International Journal of Science Education* 26, no. 5: 515–535.

Mortimer, Eduardo Fleury. 1995. "Conceptual change or conceptual profile change?" *Science & Education* 4, no. 3: 265–287.

Müller, Fritz. 2009. "Respiração nos caranguejos terrestres." In: *Para Darwin*, 57–66. Translated by Luiz Roberto Fontes and Stefano Hagen. Florianópolis: Editora UFSC.

Pauly, Daniel. 2007. *Darwin's Fishes: An Encyclopedia of Ichthyology, Ecology and Evolution*. Cambridge: Cambrige University Press.

Plato. 1887. *Meno*. Oxford: Clarendon Press.

Possani, Lourdes de Fátima Paschoaletto, Joana Alves da Silva, orgs. 2004. *Traçando o perfil de alunos e professores de EJA*. São Paulo: SME/DOT-EJA.

Prestes, Maria Elice de Brzezinski. 2003. "A biologia experimental de Lazzaro Spallanzani (1729–1799)." PhD diss., University of São Paulo.

Réaumur, René-Antoine Ferchault. 1714. "Des effets que produit le poisson apellé em François Torpille, ou Tremble, sur ceux que le touchen; Et le cause dont ils dépendent." *Mémoire de l'Academie royale des science avec les mémoires de mathématiques et de physique*: 344–362.

Redi, Francesco. 1671. *Esperienze intorno a diverse cose naturali e particolarmente a quelle che son portate dall'Indie, scritte em uma Lettera Al Padre Atanasio Chircher della Compagnia di Gesú*. Firenze: All'Insegna.

Ridley, Mark. 2006. *Evolução*. Translated by Henrique Ferreira, Luciane Passaglia e Rivo Fischer. Porto Alegre: Artmed.

Saint-Hilaire, Étienne Geoffroy. 1802. "Mémoire sur l'anatomie comparèe des organs électriques de la raie torpille, du gymnote engourdissant et du silure trembleur." *Annales du Museum National D'Histoire Naturelle*: 392–407.

São Paulo. Secretaria Municipal de Educação. Divisão de Orientação Técnica da Educação de Jovens e Adultos. 2010. *Orientações curriculares para EJA*. São Paulo: SME/DOT-EJA.

Southerland, Sherry A., Mike U. Smith, Catherine I. Cummins. 2009. "What do you mean by that? Using structured interviews to assess understanding." In Mintzes, Joel J., James H. Wandersee, Joseph D. Novak, eds. *Assessing Science Understanding a Human Constructivist View*. New York: Elsevier.

Vilanova, Rita, Isabel Martins. 2008. "Educação em Ciências e Educação de Jovens e Adultos: pela necessidade do diálogo entre campos e práticas." *Ciência & educação* 14, no. 2: 331–346.

Walsh, John, S. Siegnet. 1773–1774. "Of the electric Property of the Torpedo." *Philosophical Transactions of Royal Society of London* 63: 461–480. Accessed February 20, 2019. https://royalsocietypublishing.org/doi/pdf/10.1098/rstl.1773.0039.

Zabala, Antoni. 1908. *A prática educativa: como ensinar*. Porto Alegre: Artmed.

Chapter Seven

Challenges of Understanding Macroevolution among Brazilian Biology Students and Continuing Education Efforts

Leonardo A. Luvison Araújo, Ronaldo A. Paesi, and Voltaire Paes-Neto

Biology teaching in Brazil has been generally characterized as having a weak focus on evolutionary theory. Evolution is often covered in only a few class sessions at the end of secondary education and does not play a central role even in higher education programs (Tavares and Bobrowski 2018; Goedert 2004). Moreover, since science curricula commonly focus on microevolution (i.e., natural selection), with limited exposure to macroevolution, textbooks also fail to convey an understanding of the latter. Both Brazilian textbooks and textbooks written in English give microevolution foremost prominence when approaching evolution (Santos and El-Hani 2013; Nehm and Kampourakis 2014).

This uneven focus is not only found in Brazil, given that science curricula in other countries also emphasize microevolution (Nadelson and Southerland 2009; Nadelson and Southerland 2010; Padian 2010; Novick and Catley 2012; Nehm and Kampourakis 2014). This emphasis given to microevolutionary concepts and the subsequent disfavor of macroevolution may significantly limit student development and access to the understanding of evolution as a whole. A proper approach to its full spectrum cannot be achieved with insufficient focus on macroevolution in science education (Catley 2006).

Some researchers reported that students display difficulties in key concepts identified as essential for understanding macroevolution, like fossils, speciation, phylogenetics, deep time, and macroevolutionary patterns and processes (Nadelson and Southerland 2009; Padian 2010; Novick, Schreiber, and Catley 2014). These difficulties can be observed in students of different levels, including majors in biology.

Moreover, many of the fundamental public questions about evolution stem from macroevolutionary changes (Nadelson and Southerland 2009). Advocates of creationism and intelligent design have questioned the reality of macroevolutionary patterns due to the incompleteness of the fossil record, and suggest that evolution is a "theory in crisis" (Padian 2010; Nehm and Kampourakis 2014). Most of creationist efforts devoted to criticizing evolutionary science are denials of macroevolutionary evidence. If students are not exposed to what is known about macroevolution, proponents of creationism will be more likely to convince students that subjects like the origins of major evolutionary groups, evolutionary novelties, and adaptations are not properly explained by science.

Oliveira and Cook (2018) described the recent advancements of anti-evolution forces in the political and educational spheres of Brazil, addressing fundamental public questions about evolution that have become a matter of necessity, especially in teacher training. This is by itself a reason to give macroevolution more prominence in science education.

Thereby, this research aims to report the current situation of the understanding of macroevolution among undergraduate biology students from Southern Brazil. In response to the findings of a survey that tested students' knowledge regarding basic concepts of macroevolution, we were deliberate in incorporating content about macroevolution in a continuing education course. The authors intend to describe the course design, especially those topics related to macroevolution.

THE EVOLUTIONARY PLURALISM PERSPECTIVE

Macroevolutionary thought has a rich history, having given key contributions to our knowledge about the major changes in the form and function of organisms. Despite some variation in its definition, the fundamental idea regarding macroevolution relates to the patterns and processes that happen above the level of populational changes (Padian 2010).

There is more than one way to incorporate content about macroevolution in an education course. According to the Evolutionary Synthesis of the 1930s and 1940s, macroevolution is the result of the accumulation of differences that arise through microevolutionary processes (Laland et al. 2015; Futuyma 2015). Biologists such as Theodosius Dobzhansky (1900–1975), George Gaylord Simpson (1902–1984), and more recently, Douglas J. Futuyma have generally considered natural selection to be a sufficient causal explanation for most macroevolutionary patterns (Nehm and Kampourakis 2014; Futuyma 2015). This "extrapolationist" thesis establishes that what happens at the population level should sufficiently account for large-scale patterns in the history of life.

However, this positioning has been challenged by advances in paleobiology and evolutionary developmental theory, which urge that macroevolution is not reducible to microevolutionary processes. Evolutionary biologists such as Eldredge (1989) and Gould (1977) have adopted a causally pluralistic evolutionary worldview, which we might define as "evolutionary pluralism" (Pievani 2002). They refer to a three-dimensional pluralism: a multiplicity of rhythms in evolution, a multiplicity of evolutionary units and levels, and a multiplicity of processes in evolution. These authors do not disregard the importance of natural selection, genes, and phyletic gradualism, but have questioned their all-encompassing explanatory power.

The multiplicity of rhythms in evolution is associated with the theory of punctuated equilibrium, which predicts that a lot of evolutionary change takes place in short periods of time, tied to speciation events (Gould and Eldredge 1972; Lieberman and Eldredge 2014). In contrast, phyletic gradualism depicts evolution as a slow, steady process in which organisms change slowly over time. Pluralistic views consider both phyletic gradualism and punctuated equilibrium as valid models for understanding macroevolution. The multiplicity of evolutionary units and levels has been presented as a hierarchical theory that impinges on evolution processes: from genes, demes, and species to higher taxa (Eldredge 1985). Finally, process pluralism consists of a multiplicity of mechanisms of evolution beyond those that happen at the population level. Process pluralism in evolution invokes historical contingency, speciation processes, developmental and genetic constraints, multilevel selection, among others as important causes of macroevolutionary patterns (Gould 2002; Orzack and Forber 2017).

The pluralistic picture of evolution is no surprise, once scientific practice is struck by the multiplicity of models, theoretical approaches, and explanations for the evolutionary phenomena (Mitchell 2003). As a result, philosophers and historians of biology usually endorse some form of pluralism concerning evolution (Brigandt 2010). Scholars have distinguished different kinds of pluralism. The details of this debate, however, lie beyond the scope of this chapter (see, e.g., Mitchell 2003; Kellert, Longino, and Waters 2006). For the purposes of our discussion, we regard the several implications of pluralism for science education, especially concerning the need to teach science as an interdisciplinary and dynamic practice.

Evolutionary pluralism is the source of our own perspective for the development of a continuing education project. We endorse that macroevolution is not explained only by microevolutionary processes, but additional evolutionary processes help explain macroevolutionary patterns. Thus, the course inevitably requires knowledge from many different fields, including developmental biology, phylogeny, paleontology, theoretical biology, and ecology

(Bringandt 2010). A pluralistic approach adapted to the complexity of macroevolution is fundamental for the learning of current evolutionary theory. Consequently, the course design explicitly promotes the pluralistic agenda.

SETTING UP THE STUDY

In order to measure students' knowledge, we designed a survey that covers basic concepts of macroevolution, as has been suggested by other studies (Padian 2010; Nadelson and Southerland 2009; Gregory 2008a). Collaborators of the continuing education course (graduate students researching on evolution) examined the survey to refine and ensure its accuracy and clarity. The survey included ten items, four of which about tree thinking, a component of macroevolution that every biology student should be able to understand (Padian 2010).

The survey was applied at the beginning of each of the biannual courses in March and July of 2018. The research sample consisted of ninety biology students in their last semesters of graduation. More than 70 percent of the subjects had attended courses related to evolution before. After reading statements related to macroevolution, subjects were asked to choose among three options: *Agree*, *Disagree*, or *Unsure*. To describe the course design and measure the impact of intervention, this study uses the participant observation research method to evaluate qualitatively the continuing education efforts. The frequency of distribution of each item is shown in table 7.1.

SURVEY RESULTS

Half of the questions were correctly answered by more than 50 percent of the evaluated undergraduate biology students. It is worth mentioning that questions related to some macroevolutionary aspects, like radiations and extinctions (fourth statement) and the evidence for a common descendent to all life on Earth (sixth statement), were accurately answered by more than 80 percent of the students.

Although the percentage of students giving accurate answers to the fourth statement above surpasses that of inaccurate answers, nearly 30 percent of participants provided inaccurate answers to questions concerning evolutionary constraints (first statement), rates and patterns (second statement), and the fossil record as evidence for evolution (third statement). Moreover, the third statement is similar to an argument used by creationists and proponents of intelligent design, a concept that is critical in the context of advancements of

Table 7.1. Evolution concepts survey and percentages of participants responses.

Macroevolutionary statements	*Agree (%)*	*Disagree (%)*	*Unsure (%)*
1. Evolution is constrained by restrictions arising from development and evolutionary history of species.	38.9	27.8	33.3
2. Evolution always occurs slowly and gradually.	40	51.1	8.9
3. One problematic point of evolutionary theory is the lack of fossil records of intermediate forms, like those between humans and other primates and between mammals and reptiles.	28.9	40	31.1
4. Evolutionary radiations and extinctions are important evolutionary processes.	87.8	3.3	8.9
5. Once scientists determine that theories are well supported, they can refer to them as facts.	30	38.9	31.1
6. Eukaryotes and prokaryotes share a common ancestor.	82.2	3.3	14.5

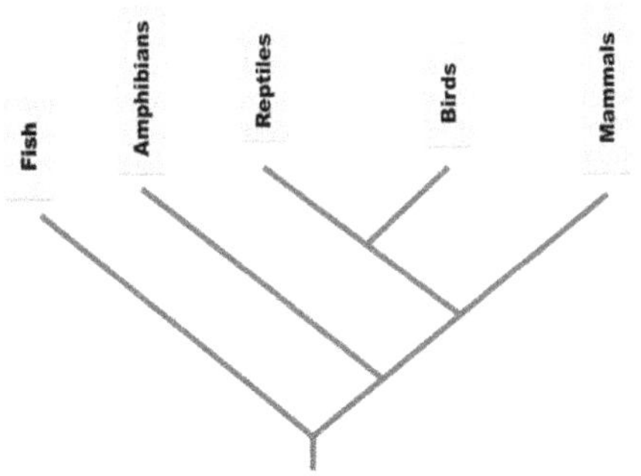

7. Birds are more closely related to reptiles than to mammals.	84.4	8.9	6.7
8. According to the phylogeny, fish represent the ancestors of modern amphibians, which, in turn, are the ancestors of modern reptiles.	25.5	60	14.5
9. Very few changes have occurred with fish since they diverged from the common ancestor of vertebrates.	32.2	43.3	24.5
10. Fish are more closely related to amphibians than to birds.	81.1	13.3	5.6

anti-evolution forces in Brazil. Similarly, the fifth statement portrays a misconception about the nature of scientific investigation that is often mentioned by advocates of creationism and intelligent design: the conflation of fact and theory (Futuyma 1983).

A large number of the evaluated undergraduate students provided inaccurate answers concerning tree thinking. Nearly 80 percent of them showed to have based their answers on the proximity of the branch tips while attempting to determine relatedness (tenth statement). They presumably did so while incorrectly interpreting the ordering of terminal nodes as meaningful, rather than assessing the pattern of branching (Gregory 2008a). Reading across the tips can also lead to incorrect intuitions about the existence of evolutionary trends among modern groups. An assumption that led to this inaccuracy is that the common ancestor of two modern groups is similar to or is one of the modern groups (Gregory 2008b). Almost 30 percent of the undergraduate biology students agreed with statements that endorse this interpretation (eighth and ninth statement).

Our results have shown a moderate comprehension of basic macroevolution concepts among participants. Unexpectedly, we have also identified a large group of students that showed to require a higher level of phylogenetic literacy in order to properly interpret evolutionary trees and, at least equally important, to be able to determine what they do not represent. The percentage of subjects to exhibit misconceptions or to lack particular skills related to tree thinking was higher if compared to what was observed in other studies involving undergraduate students (Meir et al. 2007; Kummer, Whipple, and Jensen 2016). Tree thinking is a sensitive topic within evolution teaching, and the lack of its proper understanding can be detrimental to one's knowledge about the patterns and processes that have occurred throughout the history of life. This flaw in basic training is prone to be intensified in a context of growing anti-evolution forces like the one Brazil currently undergoes.

"COURSE OF EVOLUTIONARY BIOLOGY" TEACHING INTERVENTION

Study Context and Participants

The present study highlights a continuing education project offered by a group of graduate students who conduct research on evolution at Universidade Federal do Rio Grande do Sul. Since 2014, a biannual, one-week, thirty-hour course is offered to undergraduate biology students of various higher education institutions. The course occurs in Porto Alegre and most of its participants are undergraduate biology students from public and private universities in Southern Brazil (states of Paraná, Santa Catarina and Rio Grande do Sul). Activities of the course focus on the empirical and conceptual advances in current evolutionary biology, misconceptions about evolution, nature of science, and teaching challenges.

Considering the results of the survey described above, we were deliberate in incorporating content about macroevolution and existing evolution education resources into the course design. We determined some aspects according to the literature, based on works like Nadelson and Southerland (2009), which identifies essential concepts necessary for the comprehension of macroevolution, such as deep time, phylogenetics, speciation, and fossils. In addition, we considered macroevolutionary process as an education content in fault between students, once professional background of the majority of students focus on genetic aspects of microevolution.

We organized the macroevolutionary aspects of the course in three axes: teaching tree thinking; paleobiology and museum visit; and process pluralism. The required materials were a book written by the authors of the course (Araújo 2017) and a paleontology digital book (Soares 2015). These materials are available online for free in Brazilian Portuguese.[1]

Teaching Tree Thinking

Phylogenies are increasingly important in biological research and for the way scientists perceive the natural world (Thanukos 2010). Phylogenetic analyses are used in basic science as a way to address fundamental issues related to the diversity of life and its history on our planet (Phillips et al. 2012). Since they concern hypotheses about kinship relations and changes in characters, phylogenies can be used both to describe and to understand the evolution of characters, becoming tools for testing different hypotheses involving the evolution of life (Wiley 2010).

Phylogenetic systematics has also revolutionized biological classification. All current taxonomists use the basic method of cladistics, assuming that their classifications should represent kinship relationships (Wheeler 2012). In addition to the most basic implications in evolutionary studies, phylogenies are influencing various areas of science, such as conservation biology (Crozier 1997), forensic analysis (Vogel 1998), origin of pathogens (Harper et al. 2008), and even linguistics (Atkinson and Gray 2005).

Thus, to grasp modern biology students must understand the basics of phylogenetics (Thanukos 2010). Perhaps the best way to describe the importance of phylogenetics in biology teaching has been presented by O'Hara (1997, 327): "Just as beginning students in geography need to be taught how to read maps, so beginning students in biology should be taught how to read trees and to understand what trees communicate." What O'Hara makes clear in this sentence is that students need to develop their tree thinking. This form of thinking can be defined as the ability to understand and reason from cladograms (Phillips et al. 2012).

It is important to emphasize the importance of courses that approach tree thinking, since biology undergraduate courses in Brazil rarely present the theme explicitly, with phylogenetic trees being shown along different disciplines as if it were a knowledge already possessed by the students (Lopes 2008). Tree thinking should not be understood as inherent to biology students, but as a competence that needs to be developed and practiced (Sandvik 2008). In this sense, it is not surprising that several papers show that mistakes related to phylogenetic systematics are common among majors in biology (Meir et al. 2007; Novick and Catley 2007; Sandvik 2008; Phillips et al. 2012). Among the most common misunderstandings are confusions about kinship relationships (Meir et al. 2007), which we also find in the present study (statements 8 and 10). Another common problem is the interpretation of evolutionary trees as a linear, progressive process in which the "main line" often leads to *Homo sapiens* (Gregory 2008a). This vision can be found even in current Brazilian textbooks, a fact that contributes to reinforcing anthropocentric notions among students (Paesi 2018).

All these difficulties related to tree thinking are barriers to the understanding of macroevolution patterns and processes. Therefore, to achieve a basic understanding of evolutionary trees we designed four different activities: (i) analogy of phylogenies as family trees; (ii) identifying, confronting, and clarifying misconceptions; (iii) reconstructing trees; and (iv) including fossils in trees.

Analogy of Phylogenies as Family Trees

There are some similarities between species phylogenies and human family pedigrees. The simple comparison between phylogenies and family pedigrees highlights some important points regarding the interpretation of evolutionary trees, like the observation that all living things are members of a large family tree and that their similarities and differences reflect their relatedness through common ancestors. The analogy of phylogenies with family trees is detailed in the material given to the participants (Araújo 2017, 455–472).

Identifying, Confronting, and Clarifying Misconceptions

Participants are given a brief introduction to evolutionary trees and some basic details on how they should and should not be read and interpreted. This is followed by a discussion of the most common misconceptions about evolutionary trees reported in the literature (e.g., Baum, Smith, and Donovan 2005; Gregory 2008a; Meir et al. 2007; Mead 2009). Our goal is to improve students' cladogram interpretation, including emphasis on the possibility of rotation of the branching points (nodes) without changing the relations presented.

Reconstructing Trees

We present a collection of species with discrete traits and ask students to reconstruct their evolutionary tree. This proves to be an important step for improving students' understanding of cladograms, as it has been observed by other studies (Amorim 2008; Eddy et al. 2013). We also discuss the nomenclature and concepts used to describe cladograms, among them: most recent common ancestor, monophyletic group or clade, paraphyletic group, polyphyletic group, sister groups, character, character states, apomorphy, plesiomorphy, synapomorphy, homology, homoplasy, parsimony, among others.

Including Fossils in the Trees

The use of fossils in phylogenies as an educational tool is also endorsed by the course. The inclusion of fossils among extant species gives students a perspective to understand that distinct lineages evolve independently in deep time (Gregory 2008a; Angielczyk 2009). Extinct branches (clades formed by fossil species) are good examples that current biodiversity do not represent any final progress. Comments about problematic terms, like "missing links," "living fossils," "transitional forms," and "direct ancestry" are made in order to avoid misconceptions when analyzing cladograms (Padian and Angielczyk 2007; Angielczyk 2009; Mead 2009; Torrens and Barahona 2012; Casane and Laurenti 2013).

Throughout the execution of the activities, the participants of the course are exposed to the bases of phylogenetic systematics. Important cladistics concepts are reinforced in other activities of the course, such as the visit to the paleontology museum (see later section). In addition, trees and other evolutionary representations that may express the idea of progress in evolution are discussed with the participants.

Paleobiology and Museum Visit

Fossils are a fundamental source of evidence used to support our current understanding of macroevolution. The interpretation of fossils, together with the concept of deep time, are key to understanding the patterns that emerge from the geologic record, like the evolutionary radiations and the mass extinction events. Thus, the course stressed four basic paleobiologic topics: (i) an overview of paleontology as a scientific field; (ii) a guided visit to a paleontology museum exhibit; (iii) introduction to the deep time perspective of macroevolution; and (iv) ways to use paleontology in the classroom.

Overview of Paleontology as a Scientific Field of Knowledge

We explore fossil evidence and deep time as sources of evolutionary data. Both are crucial steps for students to understand the limitations of the fossil record and put into perspective the notion of how old Earth and life are (Soares 2015; Yacobucci 2018). One of the most striking aspects of the fossil record is that it provides, for some groups of organisms, reliable information on mass extinction and evolutionary radiation events (Gregory 2008b; Benton 2015). Undergraduate students in our courses are usually not aware of the impact of biotic and abiotic events that have shaped life through time. In comments and discussions, students usually delegate a minor role for these contingency aspects while explaining current biodiversity patterns.

Guided Visit to a Paleontology Museum Exhibit

The second topic is based on a guided visit to a free exhibit on the Museum of Paleontology Irajá Damiani Pinto, whose exhibit is displayed in chronological order. During the visit, the contingent feature of evolution becomes clear through the observation of most of the important events in evolution, like the origin of multicellular organisms, the emergence of land plants, and the evolution of mammals. We provide an overview of the patterns that represent the history of life, discussing the biotic and abiotic triggers and barriers that have shaped biodiversity through time.

We also discuss museographic questions of the exhibition, like the cryptic bias toward our own evolution. Although the "anthropocentric bias" is a worldwide issue concerning museums (MacDonald and Wiley 2012; MacFadden et al. 2012; Torrens and Barahona 2012), the placement of *Homo sapiens* at the end of the visit has the potential to reflect and reinforce ideas of evolutionary progress. We discuss that this aspect of the exhibit is not strictly wrong, as there is plenty of information available on the labels and given by the museum guides. Nonetheless, the general public may misinterpret our own evolution as the evolution of life as whole. We discuss strategies to avoid this problem, like starting the tour from the present and going back toward the beginning of life or simply justifying that we are indeed talking about our own evolution.

Introduction to the Deep Time Perspective of Macroevolution

This third topic discusses themes like evolutionary rates, long-term patterns, and the emergence of higher taxa. Most of these themes are still debatable inside evolutionary biology as a scientific field (e.g., Carroll 2001; Gregory 2008b; Benton 2015). Nonetheless, we try to expose and discuss as many "scenarios" as possible. Macroevolution is sometimes understudied as the evolution of higher taxa, like phyla or kingdom, which is bound to how we classify species and understand speciation (Carroll 2001). In the course, these key macroevolutionary subjects are discussed firstly considering problems of incompatibility with the linnaean classification system (Amorim 2008; Mishler 2009).

Phylogenetic systematics is the most appropriate way to understand the relatedness of species in a system based on shared ancestry, which includes the species known from fossil records. We encourage students to adopt a cladistics approach rather than emphasizing the linnaean hierarchies. To endorse this view we compare several definitions of extant and extinct groups (e.g., grades, clades, monophyletic, crown-group, stem-group) and see how fossils may support phylogenetic studies (Briggs and Fortey 2005). When fossils are incorporated in a phylogeny, they reveal the trajectory of the acquisition of characters even between two very disparate extant groups. With such examples, the limitations of the linnaean system become clear.

For numerous scientists and philosophers, the origin, persistence, and extinction of species are also within the scope of macroevolution (Lieberman and Eldredge 2014). As illustrated by our survey, most students realize that changes within a lineage may occur at different rates (second statement). However, apparently few students realize the deep macroevolutionary component of this pattern. This is the core argument of the punctuated equilibrium theory, which is still controversial (Padian 2013; Lieberman and Eldredge 2014; Pennell et al. 2014). Despite the discussion of how important punctuated equilibrium theory can be to the evolution of life as a whole, some questions related to it are still crucial, like species stasis, evolvability, species sorting and selection, turnover events, and the relationship of disparity and diversity (Carroll 2001; Benton 2015).

Ways to Use Paleontology in the Classroom

In the fourth and last topic we encourage students to use paleontology in everyday lectures or subjects. Based on published and free resources (e.g., Soares 2015) we encourage future biology teachers to turn the deep time evolution perspective into a nondissociable dimension of every living group or ecosystem. To help students understand geologic history, we map some important events in the history of life on a timeline. Students are given a list of biological milestones to place on a geologic time scale. This assignment helps students visualize and develop their representations for deep time. This activity is detailed in the material given to the students (Soares 2015, 556–559). To understand how the fossil record operates, we use the "Tafonogame," which helps students understand how biased the fossil record can be (Soares 2015, 659–662). We also apply an activity that simulates biostratigraphic processes and demonstrates how fossils can be used by the students to infer paleoecological aspects of ancient communities.

Process Pluralism: Macroevolution beyond Natural Selection

The majority of students view natural selection and genetic aspects of microevolution as the only important causes of macroevolution. One way of dealing with this problem is to formulate a teaching approach that invokes a diverse range of evolutionary mechanisms. We propose interleaving natural selection with other evolutionary processes in order to improve the understanding of the nature of evolution on a scale at and above the level of population.

In harmony with process pluralism, we stress three themes during the course: (i) the role of developmental mechanisms, (ii) processes that influence diversity and disparity, (iii) integration of evolutionary processes and macroevolutionary patterns.

The Role of Developmental Mechanisms

Students are introduced to evolutionary developmental biology (evo-devo), which provides a set of concepts about how developmental processes influence evolution. Among the key empirical insights is the assumption that phenotypic variation involves the conservation of elements in the genetic toolkit, as well as changes in the gene regulatory machinery and its association with the timing, location, and amount of gene product (Laland et al. 2015). The genetic toolkit is important in the study of macroevolution, since major evolutionary changes do not necessarily require major genetic changes. Another related topic is the relationship between evolutionary theory and phenotypic plasticity. Environmental induction via phenotypic plasticity could eventually be genetically consolidated, leading to what has been termed as genetic assimilation or genetic accommodation (West-Eberhard 2003).

Processes that Influence Diversity and Disparity

Students are stimulated to analyze speciation and extinction patterns and their relationships with the various ecological and internal factors, such as environmental stability, climate, dispersal ability, factors related to adaptive traits, and biotic components. Some studies that infer causal relationships between ecological factors and macroevolutionary patterns are explored (e.g., Benton 2015; Lieberman, Miller, and Eldredge 2007). We also draw attention to species selection in a hierarchical perspective, indicating that selection can act at multiple levels.

Integration of Evolutionary Processes and Macroevolutionary Patterns

The connection between evolutionary processes and macroevolution is necessary to fully understand the underlying mechanisms that drive the observed patterns. A particular emphasis is given to the observation that phenotypic variation can be biased by the processes of development, which enable novel phenotypes to arise by developmental rearrangements of ancestral elements, such as in the nonrandom numbers of limbs and the parallel evolution of animal eyes, digits, and segments across a variety of taxa. We also outline great evolutionary events, such as the origin of tetrapods, the evolution of birds, the rise of dinosaurs, and many others to discuss distinct evolutionary scenarios that can generate biodiversity patterns.

A feature of the approach adopted here is a shift away from viewing macroevolution primarily in terms of the genetic aspects of microevolution. The overall structure concerning process pluralism deviates from the standard pattern of high school and college teaching, which falls into the adaptationist program, term coined by Gould and Lewontin (1979) to designate the approaches that do not recognize evolutionary processes other than natural selection. Therefore, we move macroevolutionary teaching away from the adaptationist program, as well as the assignment of causal and informational privileges to genes.

Price and Perez (2016) point out that students with conceptual understanding of a diversity of evolutionary mechanisms display fewer misconceptions about natural selection. Therefore, shifting the focus of evolution teaching to better balance natural selection with other mechanisms is a straightforward way to improve the understanding of evolution, as well as to reflect the current state of scientific research on evolution.

DISCUSSION

During the course, we have observed students' comments and difficulties regarding the activities. Many students demonstrated unfamiliarity with evolutionary processes beyond genetic aspects of microevolution. There is a tendency to reduce the presence of paleobiology and evolutionary developmental biology in preparation courses of Brazilian universities, as well as a growing importance being given to natural selection and population genetics. This perspective is rooted in historical and epistemological assumptions regarding connections between the works of Gregor Mendel and Charles Darwin (Bizzo and El-Hani 2009). Thus, throughout all the activities, aspects of paleobiology and process pluralism were the most problematic subjects among students.

On the other hand, most students involved in this study had been formally exposed to tree thinking, either through written materials, lectures, or both. Some participants' reports, together with what we have observed while applying the activities, indicate that the guided activities on phylogeny have developed the understanding of trees among participants. The analogy of phylogenies with family trees proved to be very effective in elucidating the pattern of branching, challenging some misconceptions concerning tree thinking, such as the existence of evolutionary trends among modern groups.

The learning environment in the course was purposely organized to be non-judgmental of scientific inaccuracies in students' understandings. This encouraged student active and exploratory engagement, which, when combined with an analysis of the reasoning leading up to the inaccuracies, proved to be beneficial to learning. Inviting students to reflect about their scientifically inaccurate conceptions in a positive and supporting learning environment was a productive strategy.

The possibility students had to confront their misconceptions about evolution in a friendly learning environment was viewed by participants as one of the main positive aspects of the course. The critique and suggestions offered by students referred mainly to the format of the course (held in a condensed period) and the schedule, which prevented the participation of working students.

We experienced several challenges, not just to codesign and coteach the course, but also ones related to methodological and disciplinary limitations. We utilized contributions from different biological disciplines in order to offer a more expansive perspective on the range of causal processes that may be responsible for the grand history of life on Earth. Given the diversity and complexity of contemporary biology, it is required that teachers know about the basic shape of current biological theories. Conveying to students the diversity of kinds of reasoning found in contemporary biology poses concrete pedagogical challenges. The absence of a framework that fully synthesizes development and macroevolutionary thought is a challenge in the guidance of pedagogical practices.

We hope that the results might prove sufficiently intriguing as to unveil new possibilities for later improvements in the study. How far can our course design improve students' understanding of macroevolution is still an open question. In the future, we intend to gather pre- and postintervention data to research on the effectiveness of intervention. Despite these limitations, a remarkable effect is that many students feel encouraged to continue their studies on evolution after the course. Many students return to the course and request support materials for self-learning.

In conclusion, a more pluralistic view of science and the contact with the conceptual advances in current evolutionary biology seem to motivate students to engage in learning. Future biology teachers will be at the frontlines of the public controversy surrounding the teaching of evolution in Brazil. Self-directed learning skills are a key component to deal with these challenges.

ACKNOWLEDGMENTS

The authors would like to thank all the collaborators of the continuing education course and students who took the time to complete our survey. This study would not have been possible without them.

NOTE

1. http//:editorafi.org/240biologia and https://paleontologianasala.

REFERENCES

Amorim, Dalton de Souza. 2008. "Paradigmas pré-evolucionistas, espécies ancestrais e o ensino de zoologia e botânica." *Ciência & Ambiente 36*, no. 1: 125–50.

Angielczyk, Kenneth D. 2009. "Dimetrodon is not a dinosaur: Using tree thinking to understand the ancient relatives of mammals and their evolution." *Evolution: Education and Outreach 2*, no. 2: 257–71.

Araújo, Leonardo Augusto Luvison. 2017. *Evolução biológica: da pesquisa ao ensino*. Porto Alegre: Editora Fi.

Atkinson, Quentin D., and Russell D. Gray. 2005. "Curious parallels and curious connections—phylogenetic thinking in biology and historical linguistics." *Systematic Biology 54*, no. 4: 513–26.

Baum, David A., Stacey DeWitt Smith, and Samuel Donovan. 2005. "The tree-thinking challenge." *Science 310*, no. 5750: 979–80.

Benton, Michael J. 2015. "Exploring macroevolution using modern and fossil data." *Proceedings of the Royal Society B: Biological Sciences 282*, no. 1810. https://doi.org/10.1098/rspb.2015.0569.

Bizzo, Nelio, and Charbel N. El-Hani. 2009. "Darwin and Mendel: Evolution and genetics." *Journal of Biological Education 43*, no. 3: 108–14.

Brigandt, Ingo. 2010. "Beyond reduction and pluralism: Toward an epistemology of explanatory integration in biology." *Erkenntnis 73*, no. 3: 295–311.

Briggs, Derek E. G., and Richard A. Fortey. 2005. "Wonderful strife: Systematics, stem groups, and the phylogenetic signal of the Cambrian radiation." *Paleobiology 31*, 94–112.

Carroll, Sean B. 2001. "The big picture." *Nature 409*, no. 6821: 69.

Casane, Didier, and Patrick Laurenti. 2013. "Why coelacanths are not 'living fossils': A review of molecular and morphological data." *Bioessays 35*, no. 4: 332–38.

Catley, Kefyn M. 2006. "Darwin's missing link—a novel paradigm for evolution education." *Science Education 90*, no. 5: 767–83.

Crozier, R. H. 1997. "Preserving the information content of species: Genetic diversity, phylogeny, and conservation worth." *Annual Review of Ecology and Systematics 28*, no. 1: 243–68.

Eddy, Sarah L., et al. 2013. "How should we teach tree-thinking? An experimental test of two hypotheses." *Evolution: Education and Outreach 6*, no. 13: 1–11.

Eldredge, Niles. 1985. *Unfinished synthesis: Biological hierarchies and modern evolutionary thought*. Oxford: Oxford University Press.

Eldredge, Niles. 1989. *Macroevolutionary dynamics*. New York: McGraw-Hill.

Futuyma, Douglas J. 1983. *Science on trial: The case for evolution*. New York: Pantheon Books.

Futuyma, Douglas J. 2015. "Can modern evolutionary theory explain macroevolution?" In *Macroevolution Explanation, Interpretation and Evidence*, edited by Emanuele Serrelli and Nathalie Gontier, 29–85. Dordrecht: Springer.

Goedert, Lidiane. 2004. "A formação do professor de biologia na UFSC e o ensino da evolução biológica." Unpublished master's thesis, Universidade Federal de Santa Catarina.

Gould, Stephen Jay. 1977. *Ontogeny and phylogeny*. Cambridge, MA: Harvard University Press.

Gould, Stephen Jay. 2002. *The structure of evolutionary theory*. Cambridge, MA: Harvard University Press.

Gould, Stephen Jay, and Niles Eldredge. 1972. "Punctuated equilibria: an alternative to phyletic gradualism." In *Models in paleobiology*, edited by Thomas Schopf, 82–115. San Francisco: Freeman, Cooper.

Gould, Stephen Jay, and Richard C. Lewontin. 1979. "The spandrels of San Marco and the Panglossian paradigm: a critique of the adaptationist programme." *Proceedings of the Royal Society B: Biological Sciences 205*, no. 1161: 581–98.

Gregory, T. Ryan. 2008a. "Understanding evolutionary trees." *Evolution: Education and Outreach 1*, no. 2: 121–37.

Gregory, T. Ryan. 2008b. "Evolutionary trends." *Evolution: Education and Outreach 1*, no. 3: 259–73.

Harper, Kristin N., et al. 2008. "On the origin of the treponematoses: a phylogenetic approach." *PLoS Neglected Tropical Diseases 2*, no. 1: e148. https://doi.org/10.1371/journal.pntd.0000148.

Kellert, Stephen H., Helen E. Longino, and C. Kenneth Waters. 2006. *Scientific pluralism*. Minneapolis: University of Minnesota Press.

Kummer, Tyler A., Clinton J. Whipple, and Jamie L. Jensen. 2016. "Prevalence and persistence of misconceptions in tree thinking." *Journal of Microbiology and Biology Education 17*, no. 3: 389–98.

Laland, Kevin N., et al. 2015. "The extended evolutionary synthesis: Its structure, assumptions and predictions." *Proceedings of the Royal Society B: Biological Sciences 282*, no. 1813. https://doi.org/10.1098/rspb.2015.1019.

Lieberman, Bruce S., and Niles Eldredge. 2014. "What is punctuated equilibrium? What is macroevolution? A response to Pennell et al." *Trends in Ecology and Evolution 29*, no. 4: 185–86.

Lieberman, Bruce S., William Miller, and Niles Eldredge. 2007. "Paleontological patterns, macroecological dynamics and the evolutionary process." *Evolutionary Biology 34*, no. 2: 28–48.

Lopes, Welinton Ribamar. 2008. "Ensino de Filogenia Animal: percepções de estudantes e professores e análise de propostas metodológicas." Unpublished master's thesis, Universidade Federal de Pernambuco.

MacDonald, Teresa, and Edward O. Wiley. 2012. "Communicating phylogeny: Evolutionary tree diagrams in museums." *Evolution: Education and Outreach 5*, no. 1: 14–28.

MacFadden, Bruce J., et al. 2012. "Fossil horses, orthogenesis, and communicating evolution in museums." *Evolution: Education and Outreach 5*, no. 1: 29–37.

Mead, Louise S. 2009. "Transforming our thinking about transitional forms." *Evolution: Education and Outreach 2*, no. 2: 310–14.

Meir, Eli, et al. 2007. "College students' misconceptions about evolutionary trees." *The American Biology Teacher 69*, no. 7: 71–76.

Mishler, Brent D. 2009. "Three centuries of paradigm changes in biological classification: Is the end in sight?" *Taxon 58*, no. 1: 61–67.

Mitchell, Sandra D. 2003. *Biological complexity and integrative pluralism*. Cambridge: Cambridge University Press.

Nadelson, Louis S., and Sherry A. Southerland. 2009. "Development and preliminary evaluation of the measure of understanding of macroevolution: Introducing the MUM." *The Journal of Experimental Education 78*, no. 2: 151–90.

Nadelson, Louis S., and Sherry A. Southerland. 2010. "Examining the interaction of acceptance and understanding: How does the relationship change with a focus on macroevolution?" *Evolution: Education and Outreach 3*, no. 1: 82–88.

Nehm, Ross H., and Kostas Kampourakis. 2014. "History and philosophy of science and the teaching of macroevolution." In *International Handbook of Research in History, Philosophy and Science Teaching*, edited by Michael Matthews, 401–21. Dordrecht: Springer.

Novick, Laura R., and Kefyn M. Catley. 2007. "Understanding phylogenies in biology: The influence of a Gestalt perceptual principle." *Journal of Experimental Psychology 13*, no. 4: 197–223.

Novick, Laura R., and Kefyn M. Catley. 2012. "Assessing students' understanding of macroevolution: Concerns regarding the validity of the MUM." *International Journal of Science Education 34*, no. 17: 2679–703.

Novick, Laura R., Emily G. Schreiber, and Kefyn M. Catley. 2014. "Deconstructing evolution education: The relationship between micro-and macroevolution." *Journal of Research in Science Teaching 51*, no. 6: 759–88.

O'Hara, Robert J. 1997. "Population thinking and tree thinking in systematics." *Zoologica scripta 26*, no. 4: 323–29.

Oliveira, Alandeom W., and Kristin L. Cook. 2018. "Evolution education and the rise of the creationist movement in Brazil." In *Evolution Education Around the Globe*, edited by Hasan Deniz and Lisa A. Borgerding, 119–36. Dordrecht: Springer.

Orzack, Steven Hecht, and Patrick Forber. 2017. "Adaptationism." *Stanford Encyclopedia of Philosophy*. https://plato.stanford.edu/entries/adaptationism/.

Padian, Kevin. 2010. "How to win the evolution war: teach macroevolution!" *Evolution: Education and Outreach 3*, no. 2: 206–14.

Padian, Kevin. 2013. "Correcting some common misrepresentations of evolution in textbooks and the media." *Evolution: Education and Outreach 6*, no. 1: 11.

Padian, Kevin, and Kenneth D. Angielczyk. 2007. "'Transitional forms' versus transitional features." In *Scientists Confront Intelligent Design and Creationism*, edited by Andrew J. Petto and Laurie R. Godfrey, 197–230. New York: Norton.

Paesi, Ronaldo A. 2018. "Evolução humana nos livros didáticos de Biologia: o antropocentrismo em questão." *Revista Electrónica de Enseñanza de las Ciencias 17*, no. 1: 143–66.

Pennell, Matthew W., Luke J. Harmon, and Josef C. Uyeda. 2014. "Is there room for punctuated equilibrium in macroevolution?" *Trends in Ecology and Evolution 29*, no. 1: 23–32.

Phillips, Brenda C., et al. 2012. "Teaching tree thinking to college students: It's not as easy as you think." *Evolution: Education and Outreach 5*, no. 4: 595–602.

Pievani, Telmo. 2002. "Rhapsodic evolution: Essay on exaptation and evolutionary pluralism." *World Futures: The Journal of General Evolution 59*, no. 2: 63–81.

Price, Rebecca M., and Kathryn E. Perez. 2016. "Beyond the adaptationist legacy: Updating our teaching to include a diversity of evolutionary mechanisms." *The American Biology Teacher 78*, no. 2: 101–8.

Sandvik, Hanno. 2008. "Tree thinking cannot taken for granted: Challenges for teaching phylogenetics." *Theory in Biosciences 127*, no. 1: 45–51.

Santos, Wellington Bittencourt, and Charbel N. El-Hani. 2013. A abordagem de pluralismo de processos e da Evo-Devo em livros didáticos de Biologia Evolutiva e Zoologia de Vertebrados. *Ensaio Pesquisa em Educação em Ciências 15*, no. 3: 199–216.

Soares, Marina Bento. 2015. *A paleontologia na sala de aula*. Sociedade Brasileira de Paleontologia. https://www.paleontologianasaladeaula.com/.

Tavares, Gustavo Medina, and Vera L. Bobrowski. 2018. "Integrative assessment of evolutionary theory acceptance and knowledge levels of biology undergraduate students from a Brazilian university." *International Journal of Science Education 40*, no. 4: 442–58.

Thanukos, Anastasia. 2010. "Evolutionary trees from the tabloids and beyond." *Evolution: Education and Outreach 3*, no. 4: 563–72.

Torrens, Erica., and Ana Barahona. 2012. "Why are some evolutionary trees in natural history museums prone to being misinterpreted?" *Evolution: Education and Outreach 5*, no. 1: 76–100.

Vogel, Gretchen. 1998. "HIV strain analysis debuts in murder trial." *Science 282*, no. 5390: 851–52.

Wheeler, Ward C. 2012. *Systematics: A course of lectures*. Hoboken, NJ: John Wiley & Sons.

West-Eberhard, Mary Jane. 2003. *Developmental plasticity and evolution*. Oxford: Oxford University Press.

Wiley, Edward O. 2010. "Why trees are important." *Evolution: Education and Outreach 3*, no. 4: 499–505.

Yacobucci, Margaret M. 2018. *Confronting prior conceptions in paleontology courses*. Cambridge: Cambridge University Press.

Part III

INTERNATIONAL COMPARISONS WITH BRAZIL

Chapter Eight

What Is the Role of Sound Evidence in Evolution Education?

A Research Program Following Students' Narratives in Brazil and Elsewhere

Nelio Bizzo

This chapter reports the designing of a research program at the University of São Paulo in the 1980s, which included a closer reading of Charles Darwin's *The Origin of Species* and the way it was presented to the public and taught in schools, considering the only Brazilian translation available then (Darwin 1981). No direct links between History of Science and the way students construct their knowledge today were to be found, as Piaget and Garcia (1987) had stressed. However, at that time there was a reappraisal of the need for eliciting students' ideas, this time focusing specifically on science. The project would also direct efforts to research on students' narratives about natural phenomena, and their logic, even if not always straightforward to understand at first glance.

Several sources of information available to the school community were to be studied. Richard Dawkins's famous book *The Blind Watchmaker* (1988) brought a closing chapter called "Doomed Rivals" in which he states that two "rivals"—Darwin and Lamarck—had opposing positions about two key concepts: the role of use and disuse, and the heredity of acquired characters (Dawkins 1988, pp 288–89). This opposition was called the "Darwinian Myth in the Classroom" (Bizzo and Molina 2004), which reproduces the view found in chapter 3 of the famous early *Biological Science Curriculum Study* (BSCS) textbook, blue version (see below), which also dedicated a chapter to discussing two "conflicting" points of view. This vision of the supposedly opposing views on the genetic effects of use and disuse of organs was widespread in several media, including magazines for the public, science museums and teaching materials. It was also widespread in Brazilian textbooks at least up to 2006 (Almeida and Falcão 2010).

BIOLOGY TEXTBOOKS IN BRAZIL IN THE 1980s

Most learning materials used in Brazilian high schools at that time made the assumption that there was a strong opposition between Lamarck's and Darwin's views on heredity. This account, as already mentioned, was part of BSCS materials which were initially released in Brazil and elsewhere in the early 1960s. They had been prepared after an initiative of the American Institute of Biological Sciences (AIBS), an association congregating eighty-five thousand biologists in the United States, in January 1959. A group of seventy scientists and secondary teachers took part in a meeting aimed at writing the first version of biology textbooks in mid-1960. This first preliminary output underwent piloting tests in around one hundred high schools in the United States, and in the following year happened another group meeting for the writing of the final version. To this event, several organizations in different countries were invited to send representatives, ideally a scientist and a secondary teacher. The Brazilian professor Myriam Krasilchik who was a high school biology teacher at that time and worked at IBECC, a Brazilian agency with links with UNESCO, wrote about the work, which was carried out together with Professor Oswaldo Frota-Pessoa:

> The IBECC was invited to join the project of elaboration of the Biological Science Curriculum Study (BSCS) sending two representatives, a university professor and a high school teacher, to take part in the formulation of project of Biology teaching, as it was usual with the American group, in which a scientist worked together with a high school teacher. (. . .) at the University of Colorado at Boulder, where the team in charge of the first version of the BSCS was based. (Krasilchik 2012, 201–2)

As part of that group, Professor Krasilchik met very famous scientists, such as Fritz Went, John Moore, Bentley Glass, Marston Bates, Joseph J. Schwab, and Arnold and Hulda Grobman. The first generation of BSCS books had three versions, and two of them (Blue and Green) were published in Brazil. The Blue Version [*Biological Science: Molecules to Man—Biological Science Curriculum Study (BSCS) Blue Version*] was published in Brazil as a preliminary edition entitled *Biologia: Das Moléculas ao Homem* (Rio de Janeiro: Centro de Publicações Técnicas da Aliança, 1965). The green version was published years later, as *BSCS, Biologia: Versão Verde* (São Paulo: Edart, 1972), both with two volumes for the students. These books brought particular stress on History of Science (HS) and were not merely translations as some local contexts were added:

> The history of Biology and the process of scientific inquiry were essential innovations in the presentation of this science (. . .) The BSCS books were very

different from others being used at that time, and this brought difficulties to their adoption, but introducing the molecular level at the beginning of the course was an outstanding feature and influenced the posterior elaboration of curricula and other didactic materials, provoking a significant change, bringing the term DNA as a universal expression (Krasilchik 2012, 203).

Although BSCS Brazilian publications cannot be considered real bestselling books, they had a consistent sales success, after a difficult beginning. As no local publisher was committed to invest, the United States Agency for International Development (USAID) financed a preliminary edition, followed by an agreement between IBECC and the University of Brasília in 1965 (Barra and Lorentz 1986). A private publisher (EDART) was responsible for the subsequent releases, which would encompass the green version. The blue version had at least seventeen printings from 1965 to 1977. From 1965 to 1972, some 224,000 books of the blue version were sold in Brazil (Barra and Lorenz 1986). The BSCS books were very influential in the country, as other textbooks adopted some of the new approaches they brought, particularly the blue version, in terms of molecular approach. More importantly, teacher training courses were organized on a large scale and involved a high number of science teachers in different regions of Brazil, providing direct contact with the textbooks (Barra & Lorentz 1986; Krasilchik 2012).

A careful study of the biology textbooks produced in Brazil in 1968–1978 showed that five new productions in the period brought a very similar approach to that of the BSCS blue version (Almeida and Falcão 2010).

THE VOICE OF THE STUDENTS

Eliciting students' thoughts about evolution was not yet part of the common academic research practice, as it would become soon after, in the early 1990s. The first academic work dealing with the subject, as far as we could figure out, was John Deadman's PhD work presented to Chelsea College (University of London), in 1976. He had interviewed fifty-two students who had not been taught the subjects of genetics and evolution at a grammar school near London (Hertfordshire), where we worked as science teachers. Students belonged to secondary school, were boys only, aged eleven to fifteen. One year later, almost all the students (n = 49) were interviewed again, with a particular focus on two aspects: the causes/reasons of evolution, and the processes of biological change (Deadman 1976; Deadman and Kelly 1978).

The findings of this seminal work included the categorization of three basic patterns of thought about the processes of evolution. Some students described evolution as something which had happened in the past, as part of an episodic happening. Another group of students could perceive some continuity between past and present time, but another group could see it clearly. Younger students tended to be in the first group, whereas the eldest concentrated in the last group, which was the largest one, with thirty-four out of forty-nine students (Deadman, 1976).

It is remarkable that there were no references to the question of religion (creation/evolution), showing a marked distinction between British and American educational environments at the time. The Brazilian atmosphere did not indicate that this issue could play any significant role, contrary to later developments. For this reason, the British pieces of research on the subject seemed to be more relevant to the progress of this research in Brazil.

Some texts aimed at the public, as mentioned in Dawkins's example cited above, and biology textbooks stressed the difference between Lamarck and Darwin, in terms of hereditary effects of organic changes due to use/disuse. However, direct contact with *Origin of Species* could easily show Darwin's firm belief in the inherited effects of such changes:

> Habit also has a decided influence, as in the period of flowering with plants when transported from one climate to another. In animals it has a more marked effect; for instance, *I find in the domestic duck that the bones of the wing weigh less and the bones of the leg more, in proportion to the whole skeleton, than do the same bones in the wild-duck; and I presume that this change may be safely attributed to the domestic duck flying much less, and walking more, than its wild parent*. The great and inherited development of the udders in cows and goats in countries where they are habitually milked, in comparison with the state of these organs in other countries, is another instance of the effect of use. Not a single domestic animal can be named which has not in some country drooping ears; and the view suggested by some authors, that the drooping is due to the disuse of the muscles of the ear, from the animals not being much alarmed by danger, seems probable. (Darwin 1859, 11)

The issue of inheritance of acquired characters was a central point in designing the piece of research aimed at eliciting students' conceptions of evolution. The above famous passage of Darwin's book was chosen to check whether students would agree with the inherited effects of habits or not.

It is interesting that the only Brazilian translation of Darwin's famous book available at that time (Darwin 1981) had some mistakes in this precise paragraph. Instead of mentioning wild and domestic ducks ("patos"), it mentioned wild and domestic "canaries" ("canários"), the small singing bird. In the same phrase, instead of mentioning the effects on the bones of the legs ("pernas"), it mentioned thighs ("coxas"), hence conveying an awkward image, once the thigh has just one bone (femur). It is probable that these mistakes derived, at least partially, from the use of a French version translated by Mme. Clemence Royer. She used the word "canard" for duck but strangely, contrary to the 1873 translation of J. J. Mouliné, and she chose the word "cuisse" (thighs) instead of "jambe" (leg). Due to these problems with the French and Brazilian translations, this specific example was then chosen to test Brazilian students' ideas in a written test, since it was unlikely that students were aware of Darwin's original terms on this specific issue. The mentioned chapter 3 of the Brazilian BSCS blue version brought a bibliographical suggestion for further reading in Portuguese which brought exactly the same mistake (Darwin 1946). In conclusion, it was likely that even the students' teachers were unaware of those original terms.

The final draft of the planned work included interviews with a dozen high school students who had already been taught the topics of genetics and evolution, and some two hundred students would answer a short written test. Besides, some additional studies with HS, including access to primary sources available in England should be done, aimed at further understanding the growth of the evolutionary theories in Darwin's written documents and books. There was a definite option toward a combined quantitative/qualitative research methods, following recent works recently published then (Smith and Heshusius 1986).

These research studies were carried out between 1989 and 1991, including three high schools in São Paulo, with students from several religious profiles. It also included a period of research on HS in England, with access to the Manuscripts Room, University of Cambridge Library (Cambridge), British Library (London) and Down House (Downe, Bromley).

FIRST FINDINGS OF BRAZILIAN EVOLUTION EDUCATION

The written test mentioned above was composed of five questions, with a particular focus on the views of processes involved in the generation of diversity and heredity. Students had to choose an option and write a few lines. In the first question, a small text described the fact that wild ducks can fly very well and have heavier wing bones and lighter leg bones compared with domestic ducks, which are unable to fly. They were asked to choose one explanation for this fact and write down some lines justifying the choice. One text followed Darwin's view (see above, Darwin 1859, 11) and the other presented the differences as random variations, which would have been selected by farmers in the past. Only 13 students out of 192 (or 7 percent of the answers) answered accurately mentioning the effects of random variations selected by breeders. Another question asked students their opinion about the possible origin of blindness in fish living in dark caves. A total of 123 out of 192 answers (64 percent) credited the blindness to lack of light or disuse of the eyes.

As one of the schools belonged to the Jewish community, and Darwin had expressed views about the traditional practice of circumcision, it was feasible to include one question on the subject. Darwin had written: "Blumenbach, however, asserts that Jews are often born in Germany in a condition rendering circumcision difficult so that a name is given them signifying 'born circumcised'" (Darwin 1875, 467). We then asked students to write down their opinions about the same statement, omitting the author of the information, simply stating "some physicians have reported . . ." Some answers included phrases like: "removing the DNA of that part for so many generations . . ." Only 24 percent of the students answered that there should be no relation between the two facts, with some sort of assertion contrary to the hereditary effects of amputations.

Results showed that Darwin's original views on heredity were widely shared by that group of students (Bizzo 1991, 1994). The picture was very similar to the results obtained with university students from Michigan University (Bishop and Anderson 1990).

Additionally, eleven in-depth interviews were done in the schools' environment, with no communication among the students involved. They were fully transcribed and presented some common aspects. One of the first questions was: "What do you know about biological evolution?" Students, with no exception, answered by mentioning the evolution of humans from apes, in a clear indication that human evolution was a central reference for biological change (Bizzo 1991). Thus, the following general trends in students' ideas were manifest in our results:

- The environment is the main factor for the generation of inherited variation
- The adaptation was believed to be an individual process of adjustment
- Evolution is related to progress, improvement, and growth
- Biological and cultural evolution is not considered contrasting processes
- Models for evolution were co-constructed with those for reproduction and heredity

Our empirical results confirmed what had been recently published at that time about the interconnection between the construction of ideas on evolution and heredity (Kargdo, Hobbs, and Erickson 1980; Clough and Wood-Robinson 1985). These results emphasized the need for innovation in the research field, as many researchers were claiming for more focus on teachers' professional skills, on the one hand, and, on the other, on students' cognitive abilities. Construction of knowledge in the classroom environment was part of a broader sociocultural perspective (Wertsch 1985). These results led to a new focus in the subsequent work in two manners. First, the relation between teaching Darwinian evolution and Mendelian genetics should be revised on HS grounds, and, second, a new hypothesis for historiographical research was possible.

A NEW HYPOTHESIS FOR HISTORIOGRAPHICAL RESEARCH

An Evolution Education Research Conference gathered forty-six scientists, science teacher educators and science teachers at Louisiana State University campus at the beginning of December 1992. The conclusion of the conference outlined a research agenda with seven themes, focusing mainly on students' cognitive learning potentialities and teachers' teaching abilities. There was no explicit mention of a need for new interpretations of the HS to inform the organization of content and curricula (Good et al. 1992). However, contradictions between the educational accounts of Charles Darwin's ideas against his own written words offered clues that some relevant work had yet to be done.

Students' narratives showed how important the figure of human species was in their minds. Students organized their knowledge about biological evolution around a very central idea related to human evolution (Bizzo 1991, 1994). This anthropocentric perspective was surprising as students were not taught the topic of evolution in the context of the origin of modern humans. This issue was (and still is) relevant as it might bring some real consequences for curriculum design as humans were not addressed at the beginning of a traditional exposition of the development of the theory of evolution at the high school level in Brazil.

HS scholars had no consensus on how central human species was in the so-called Darwinian revolution, at least when *The Origin of Species* was first published (1859). Historians were divided about the absence of human examples in *The Origin of Species*, as different ideas prevailed at the time. One of them argued that Darwin had encompassed humankind in his theorizing (Bajema 1988). A slightly different view maintained that he had decided not to be explicit about human evolution (Cooke 1990). Another aspect asserted that Darwin's references to humans, particularly concerning cultural evolution, did not mean that he was giving an evolutionary account of human origins. This issue he would have purposely decided to avoid when writing *Origin of Species* (Bowler 1989).

Furthermore, even though there was a vivid debate about the inclusion of humankind in the *Origin of Species*, the version which has prevailed in the educational context was linked to Galapagos finches and the like, but not to human origins. There was no mention of human evolution in the famous chapter 3 of the BSCS Blue version, which had been, as said before, very relevant in shaping the presentation of evolution in Brazilian textbooks in the 1980s.

The centrality of human evolution was of paramount importance in the proposed heuristic tool, generating a hypothesis for the reappraisal of the position of humankind in the base of the evolutionary theorizing. The first results, after analyzing the changes introduced in the several editions of *Origin of Species*, revealed interesting points to be considered. The sixth (and last) edition (1872), considered as the "canonical" version for translation purposes, does not bring an important statement about the differences of "man races," when discussing sexual selection:

> I might have adduced for this same purpose the differences between the races of man, which are so strongly marked; I may add that some little light can apparently be thrown on the origin of these differences, chiefly through sexual selection of a particular kind, but without here entering on copious details my reasoning would appear frivolous. (Darwin 1859, 199)

As this part of the paragraph was dropped in the 1872 edition, many readers around the world were unaware of this brief thought, which expressed undisputed views of Darwin's original theorizing. Others, knowing the dynamics of the text, may have concluded that there was a change in his positions about the subject, without considering that a long argument on the issue had been published in full the previous year (*Descent of Man*, 1871, 2 vols.). Therefore, keeping this part unaltered in the last edition may have been considered a neglected editorial detail. The 1981 Brazilian translation of "Origin of Species" corresponds to the last edition and does not bring this reference, as well as the Portuguese suggested reading of the mentioned Brazilian translation of BSCS.

Further research in primary historiographical sources in several places in the United Kingdom showed that Darwin had been researching the topic of "human races." He was collecting material for his "Big Species Book," which he was writing in the 1850s, as we can conclude by the marginalia penciled in some of his books. He not only possessed the last edition of the famous work by James Cowles Prichard ("Researches into the Physical History of Mankind," 3rd ed., 1841, 3 vols.) but also had worked together with him soon after realizing the power of natural selection. They both took part in a committee created by the British Association for the Advancement of Science in 1839, aimed at gathering information on human populations around the world and eventually led by the famous physician Thomas Hodgkin (Lester and Dussart 2014).

There was a vibrant discussion among historians about what was in Darwin's mind when writing *Origin of Species*. The conclusion about the specific debate supported the view that the book was primarily intended to deal with humans. The known quote on the origin of man through sexual selection, which was understandably dropped, and the famous phrase "light will be thrown on the origin of man and his history," which resisted all editions, could be taken as indications of his will to include humankind in his evolutionary views. However, some additional facts were not considered in the issue of considering humankind in the evolutionary theorizing. The reason for his ideas on natural and sexual selection applied to the "races of man" did not appear clearly in his 1859 book seemed to be due to chance factors and to a dramatic irony, related to his conceptions of a phenomenon which was called "reversion to the wild state" (Bizzo 1992b).

Students tend to co-construct their conceptions of evolution and heredity jointly. Some reports of HS state that there were no such parallels in the original construction of the theory, as Darwin "lacked" a theory of heredity. The centrality of inheritance and reproduction in students' narratives led to a reappraisal of the development of Darwin's ideas on heredity, and how (in) compatible they could have been with Mendel's. The English thinker attributed an essential role to the environment in the origin of variations, modifying hereditary particles ("gemmules"), a model which was called "soft inheritance." Again, it was possible to figure out an open field for HS research, with real impact on the issue of the structuring of curricula worldwide (Bizzo 1991; Bizzo and El-Hani 2009)

The findings of this part of the research program brought interesting new perspectives for the interconnection between science historians and educators. On one hand, new versions of HS can convince educators that alternative ways of thinking held by their students are not in fact shared by famous scientists from the past. On the other hand, educational research can provide a useful working hypothesis for the historiographical reappraisal of central parts of the development of biological sciences. This "bottom-up approach" (Bizzo 2011; 2013) can grant new clues for further inquiries on HS relevant specifically for biological education.

SHARED MEANINGS ABOUT GENETICS AND EVOLUTION

After some years of work, there was no doubt that the view of the inheritance of acquired characters had been deeply rooted in our culture. Students did not learn misconceptions about the effects of the environment in offspring by simply reading their textbooks or science magazines. So, a new phase in the research project was designed. This new phase aimed at understanding the social and historical roots of the knowledge which was shared by so many students in different countries. A new theoretical thrust was needed, with a broader focus, avoiding too much attention on the development of individual learning (or teaching) skills of the Piagetian tradition.

Developmental psychology provided relevant inputs, especially with the then-innovative ideas of sociocultural studies, from authors such as Barbara Rogoff, Jean Lave, and Etienne Wenger. Studies in the sociology of knowledge were also very useful, providing theoretical tools to understand the social construction of the so-called "common-sense knowledge" that constitutes the reality of everyday life for the ordinary member of society. By mid-1990s the expression "misconceptions industry" was well known within the science education community, and a new approach was needed, in order to understand social knowledge as a composition of different "zones of familiarity," in which some areas of reality are more illuminated than others (Berger and Luckmann 1966). The search for such "zones" led to the search for social groups with privileged access to appealing evidence. People with direct contact to sound evidence about heredity and evolution were invited to express their views, and their particular ways of conceiving reality.

One of the first steps of this new phase was to follow families with genetic syndromes as part of a collaborative effort with the group of Genetic Counseling Service (GCS), Laboratory of Human Genetics, Department of Biology, Biosciences Institute, University of São Paulo.

The first results showed a very complex web of relations, as research work occurred far away from schools, and outside the school environment which was familiar to us. Patients, and not students, were to be approached, what brought an entirely new universe of challenges related to the psychological aspects of genetic counseling (Kesser 1989).

In the sociocultural theoretical perspective, cognitive activity is not a single and purely logical mental task, but rather a two-leveled process which makes use, on the one hand, of tools for cognitive activity (such as scientific theories) and practices socially negotiated. On the other hand, immediate social interaction creates a material basis for individual cognitive action, such as explanations of heredity in order to explain the reappearance within a particular family of a malformation in a newborn, or conceptions of petrified bones, teeth and shells, within a community living in a rich fossiliferous region, as we will see ahead.

One remarkable aspect was the fact that families arrived at the university GCS after having spent long periods of time following physicians' advice, and listening to explanations about how deafness could result after some viral infections during pregnancy, and how albinism could be transmitted from one generation to another. To find what we called "first-hand discourses," spontaneous thoughts of shared meanings, we looked for families with rare syndromes, preferably large families with disorders of specified complexity, with incomplete (or reduced) penetrance and variable expressivity. A pilot study was then carried out after obtaining permission to contact relatives of one family under counseling which fulfilled the requirements of complexity. The family had a rare syndrome with a wide range of pathologies, ranging from tibial malformation to complete absence of the bone. This family also had ectrodactyly, an abnormality with several degrees of severity, involving partial to total absence of some bones in the hand. The hereditary pattern of these abnormalities is still unknown; therefore, members of the family exchanged ideas about heredity with no significant influence of physicians or GCS staff.

Our work involved the development of methodological procedures which would allow capturing meanings that members of the family transmit by their discursive activity. The main objective was to understand how members of a family highly motivated to think about heredity conceived and negotiated their views of the relationship between genetic and environmental determinants of body characters. Religious beliefs, moral, and supernatural punishments, but also unrealistic "accidents" for missing bones in the hand were mentioned as causal factors for the appearance of abnormalities and malformations. The system of shared meanings indicated that they would have turned into inheritable characters for future generations.

The decision of no publication of the interviews was grounded on the lack of proper methodology dealing with such sensitive details; a mother describing her thoughts about her baby with a specific malformation could not be analyzed with the same tools as could be done with students representing their thoughts about evolution. Results of this new phase helped to foster methodological procedures about the selection of contexts and data collection and analysis of discursive activity within the theoretical thrust of sociocultural studies.

Another research study followed, with interviews of 100 members of two large inbred families, with a common origin in Serrinha dos Pintos (RN), northeastern Brazil. The largest group consisted of sixty people, with twenty-five affected by a previously unrecognized neurological disorder. Major manifestations included congenital optic atrophy, early-onset progressive spastic paraplegia, and acoustic startle. The other group consisted of forty people with neurofibromatosis type 1 (NF1). Both groups fulfilled the requirements of complexity, and narratives were vibrant, with the environment playing a significant role to explain disorders and birthmarks, which would have turned out to be inherited. A groundbreaking work, this research not only refined the new methodology for data collection but also allowed one to understand the vital role of the influence of the environment as a causal factor to conceive genetic traits in inbred families. The syncretism of explanations incorporated scientific and lay knowledge, creating narratives which were legitimate for the members of the community. Communication of results was planned to have a central commitment to respect for the families involved (Santos, 2003; Santos and Bizzo 2005; Santos 2006). Eventually, the neurodegenerative disorder was described as a new genetic syndrome: Spastic Paraplegia, Optic Atrophy and Neuropathy (SPOAN) Syndrome (Macedo-Souza et al. 2005).

After some years of hard work, the sociocultural roots of the belief of inherited effects of acquired characters were well documented, with a solid empirical base. This knowledge is a result of a process of legitimation, which is not only a normative element but also a cognitive one. "In other words, legitimation is not simply a matter of 'values.' It always implies knowledge as well" (Berger and Luckmann 1966, 86). Interestingly, Charles Darwin was also a member of an inbred family and had had two children who probably were born with some genetic disorders (Bizzo 1992b).

LIVING NEAR FOSSILS

The above set of studies paved the way for further research, now focusing highly motivated people with direct contact with shreds of evidence of evolution, which were relevant in HS. Data collection of discursive activity about evolutionary phenomena involved young students who lived in places with plenty of fossils, looking for ways in which the community conceived and represented animals and plants from the past.

The first place that drew our attention was the village of Villavicencio, in the Andes, where Darwin reported his astonishment after finding compelling geological evidence which would shake the grounds of the "theory of the formation of the world," as he wrote to his sister Susan:

> In these same beds (and close to a Gold mine) I found a clump of petrified trees, standing upright, with the layers of fine Sandstone deposited round them, bearing the impression of their bark. These trees are covered by other Sandstones & streams of Lava to the thickness of several thousand feet. These rocks have been deposited beneath water, yet it is clear the spot where the trees grew, must once have been above the level of the sea, so that it is certain the land must have been depressed by at least as many thousand feet, as the superincumbent subaqueous deposits are thick.— But I am afraid you will tell me, I am prosy with my geological descriptions & theories.— (Darwin's letter no. DCP-LETT-275).

Applying the new methodology on discursive activity in the very same settings seemed to be an exciting challenge. The main objective was to listen to the people living near this "clump of petrified trees," explore their impressions on the marks of the trees' barks, and examine what they thought about the amount of time required to transport those remains to the height of twelve thousand feet.

Unfortunately, after a great deal of work for funding, planning, and implementing several travels, results were disappointing, as the place was desert, hardly known by local people living either Villavicencio or Uspallata (the nearest villages). An article (Bizzo and Bizzo 2006), a book (Bizzo 2009) and a TV documentary *Darwin nos Andes* (available at http://neliobizzo.pro.br/midia/documentario-darwin-nos-andes/) were the only concrete results of this endeavor. As the place is close to the country border, the nearest people living there are members of the Argentinian Armed Forces, who stay in the place for short periods.

A similar place, highly populated, was identified in northeastern Brazil, where there are significant fossil remains from the Cretaceous, in the border Pernambuco-Ceará, in northeast Brazil, a place called Chapada do Araripe. It has been mined commercially for cement manufacture and paving stones. In the quarries where the slab limestone is extracted a massive number of rare and beautiful fossils occur, rendering world-class status to the site. It constituted what is known as a "Fossil Konservat Lagerstätte," a deposit known for the exceptional preservation of fossilized organisms, where the soft parts are preserved in the form of impressions or casts. It proved to be "one of the most diverse assemblages of a Cretaceous terrestrial biota known elsewhere" (Martill et al. 2007, 4).

As a result of our research in that place, we discovered unnoticed evidence of social views about geological time in the local culture. Artisans sold paving stones with engraved images of the fossils. Contrary to the first impression, which would present them as mere counterfeit souvenirs intended to persuade gullible tourists, the carved images were artistic representations which reproduced the artisans' perceptions of the meaning of fossils. In the artisans' view, they were valuable for their intrinsic aesthetic meaning. These engravings were sold at the home of the own artisan, who was proud to say that they are handmade, showing images not only of locally known fossils, such as scorpions, dragonflies, and insects but also Jurassic Park's dinosaurs and even parrots (Figure 8.1).

These souvenirs expressed the natives' understanding of the meaning of fossils, not as proofs of what life was about one hundred million years ago in high Crctaccous, but rather as beautiful forms carved in solid stone. The fossilized feather found in the artifacts, which was on display in the local pa-

Figure 8.1. Artisans sold paving stones with engraved images of fossils; including scorpions, dragonflies, insects, parrots, and even dinosaurs.
Photo provided courtesy of Nelio Bizzo.

leontology museum as a tantalizing fact, was not understood as a proof of the antiquity of birds in the region (and the whole continent), but as a present-day parrot (figure 8.1, right-hand side).

Seven students who lived in Araripina (PE), aged nine to thirteen, and eleven students from Santana do Cariri, aged eleven to fifteen were interviewed and it was possible to capture in students' discourse is part of a social system of shared meanings for fossil evidence. Fossils were seen as signs of present-day animals and plants, which would have undergone fossilization the previous year, during the dry season. The names for fossil animals correspond to those attributed to the living ones, such as "piabinha," a small fish which can be seen in the small rivers or fossilized in limestone.

Another group of nine students, aged eight to twelve, were interviewed in the extreme south of Brazil, in a small town called Mata (RS). The place is well known for the immense quantity of fossils, mainly trunks from the Upper Triassic. Results were exciting, with rich narratives combining elements from the local culture, religious beliefs, and scientific information provided by local museums and schools. The idea of geological time proved to be very difficult. For instance, students could admit that some trees had died some two hundred million years ago, but they believed local natives cut them. They were seen as present-day local trees, known as "araucárias." In conclusion, these places had fossils very similar to present day living beings, and the idea of extinction was not evident in students' narratives (Oliveira 2006, Bizzo and Oliveira 2011).

Another set of interviews was carried out in Italy. For this part of the research, the region of Verona was chosen for several reasons. There were many indications that the area was an essential context of the first discussions about what was to be called "geology" in the eighteenth century, including the origins of mountains, their shapes, and relations that could be established with sacred books. The places selected to conduct interviews were very unique as one can easily find fossils relatively recent (fifty million years), but different from living beings found there now.

The villages of Bolca and Roncà were the two places were students were interviewed. These places bear fossils of tropical animals and plants, which could never live in the same area now with a severe winter. The small village of Roncà have fossils of manatees, Bolca have fossils of tropical fish and fruit, such as coconuts! The striking difference between the mineralized remains and the living beings of the place creates a situation which requires a special effort to conceive such contrasts. We called these places "paleoheteromorphic," contrary to the previous ones, which we called "paleoisomorphic" (Bizzo 2011; 2013; Bizzo and Oliveira 2011.

Twelve students aged nine to twelve years old took part in the interviews, which were carried out individually, in students' own homes or otherwise in places suitable for them. Interviews were performed before their parents; however, with no direct eye contact so that the child could not receive positive/negative feedback from his/her parent, who signed a written agreement. They were performed in Italian (and not in Veneto dialect), which is considered the proper form to communicate at schools, and were tape-recorded, translated into Portuguese and transcribed or summarized for analysis.

Italian children had a good idea of the names of the fossilized creatures, and their narratives are rich in details, connecting geologic and paleontological information and terminology. They know local museums well, which are visited every year. Geological time is hardly understood, but they know well that those creatures are extinct and no longer can be found anywhere else in the world.

We found significant differences in the narratives of these two group of students, even though geological time was equally hardly perceived, even when students were well aware of the paleontological terminology. Therefore, the social construction of the idea of geological time is not expected even in places with plenty of evidence. Fossils were regarded as remains of living beings, also though biblical deluge was referred by some students who belonged to both Catholic and Evangelic families. The idea of extinction was not clear for the Brazilian children, who lived in "paleoisomorphic" places, whereas the contrary was found with the Italian children, who lived in "paleoheteromorphic" places (Bizzo 2011, 2012; 2013; Bizzo and Oliveira 2011, 2012). These results are relevant not only for educational purposes but also for conservation ends.

Interestingly, the fossils of Bolca were at the center of a dispute about the possibility of extinction by the end of the eighteenth century. Creationist priests, such as Domenico Testa and Serafino Volta insisted that they belonged to living species, whereas some of the first modern geologists of the time, such as Alberto Fortis and Giambattista Brocchi contested their conclusions. The issue was solved, surprisingly, by the young Louis Agassiz in 1833, who concluded they were all extinct; in other words, they were all "paleoheteromorfies" (Bizzo and Oliveira 2012). However, when Charles Darwin was writing his famous *Origin of Species* the issue of extinction was still under discussion, not only from the perspective of the extent (local vs. global) but also from the perspective of reversion. He mentioned the famous "Barrande's colonies" as phenomena yet to be explained, and the French paleontologist replied with a new book questioning Darwin's statements soon after.

CONCLUDING REMARKS

For readers of Charles Darwin's original works and personal letters, it was clear that new sources of information were needed for the Brazilian students (and biology teachers) at least since 1981. The work on HS has proved to be crucial as a source of creative ways to conceive of available information. Charles Darwin did not discover any tantalizing fact, animal or plant, but instead ascribed meaning to available evidence in highly original and revolutionary ways. The research work reported here shows how important is the deep understanding of these past creative processes and how educators can play an important role to accomplish this task today.

The creative thinking of past scientists has some similarities with the challenges that young students face at present. It is well possible to develop new theories for historiographical research after listening to what people think about the key features of certain phenomena especially when they have direct access to the events under study. It was remarkable how some young students attributed centrality to human evolution and had a "Lamarckian" model to explain the evolutionary change.

The narratives of members of inbred families with genetic disorders bought a new perspective in order to understand how central the role of the inheritance of acquired characters is to keep up with genetic evidence. The recent research on epigenetic phenomena will possibly enhance this tendency in the near future.

Young people living in places with plenty of fossils have narratives, which allow a broader understanding of different ways in which evidence regarding evolution can be conceived. The concept of deep time cannot be expected to appear spontaneously, and therefore must be explicitly addressed by curriculum development policies. The concept of "paleoheteromorphies" played an essential role for the understanding of the social construction of the idea of extinction, which seems to depend to a certain extent on the recognition of the differences between past and present-day biotas.

This research program is still going on, now with a new branch related to international and nationwide surveys. Further studies try to understand the level of acceptance of science, technology, and evolution among students at the high school level, painting a more comprehensive picture (see next chapter). HS studies on Italian geology of the eighteenth to nineteenth centuries are also presently underway focusing on the obstacles faced by the burgeoning Italian geology community (Bizzo 2011; Bizzo and Oliveira 2012). The idea of irreversible extinctions is a critical concept in *Origin of Species*, and was still under discussion at that time. Darwin seems to have been influenced by the ideas of Giambattista Brocchi since his first observations during the voyage of the Beagle (Dominici, 2010, Dominici and Eldredge, 2010). The similarities between key concepts hard to develop in the classroom and HS episodes suggest it is worthwhile to make such texts available to educators.

ACKNOWLEDGMENTS

Several people and institutions contributed to the funding and implementation of the many research studies reported here and deserve special thanks: Myriam Krasilchik, Wynne Harlen, Peter Kelly, David Hamilton, Terry Russell, Marsha Richmond, Heidi Bradshaw, Oliver Harlen, Edgar Jenkins, Jonathan Hodge, Greg Radick, Everett Mendelsohn, Luca Ciancio, and Roberto Zorzin. A great deal of help and enthusiasm were received from the late professors Oswaldo Frota-Pessoa, Bernardo Beiguelman, Francisco Salzano, Newton Freire-Maia, and Terezinha Collichio. Special thanks to University of São Paulo (Pró-Reitoria de Pesquisa, EDEVO-Darwin), Liverpool University, Cambridge University Library (Manuscripts Room and Rare Books Room), British Library and Royal College of Surgeons of England (formerly responsible for Down House access), Leeds University, Università degli Studi di Verona, Museo di Storia Naturale di Verona, and Università degli Studi di Padova. Funding agencies include Interamerican Development Bank (USP/BID), MEC/CAPES (proc. 3955/92-16) CNPq (proc. 300786/93-0, 304243/2005-1, 302107/2008-8, 308899/2011-3, 308877/2015-2), and FAPESP (proc. 92/01278-4, 96/04547-7, 00/03660-1, 2010/10320-4, 2016/05843-4).

REFERENCES

Almeida, Argus V. D. and Jorge T. Falcão. "As teorias de Lamarck e Darwin nos livros didáticos de Biologia no Brasil." *Ciência & Educação 16*, no. 3 (September 2010): 649–665.

Bajema, Carl. "Charles Darwin on Man in the First Edition of Origin of Species." *Journal of the History of Biology 21*, no. 3 (March 1988): 403–410.

Barra, Vilma. M. and Karl M. Lorentz. "Produção de materiais didáticos de Ciências no Brasil, período: 1950 a 1980." *Ciência e Cultura 38*, no. 12 (December 1986): 1970–1983.

Berger, Peter. L. and Thomas Luckmann. *The social construction of reality*. New York: Doubleday & Co., 1966.

Bishop, Beth. A. and Charles W. Anderson. "Student conceptions of natural selection and its role in evolution." *Journal of Research in Science Teaching 27*, no. 5 (May 1990): 415–427.

Bizzo, Nelio. "Ensino de Evolução e História do Darwinismo." PhD thesis, University of São Paulo, 1991.

Bizzo, Nelio. "História da Ciência e Ensino: onde terminam os paralelos possíveis?" *Em Aberto 11*, no. 55 (July/September 1992a): 29–35.

Bizzo, Nelio. "Darwin on Man in Origin of Species: Further Factors Considered." *Journal of the History of Biology 25*, no. 1 (March 1992b): 137–147.

Bizzo, Nelio. "From Down House Landlord to Brazilian High School Students: What has happened to evolutionary knowledge on the way?" *Journal of Research in Science Teaching 31*, no. 5 (May 1994): 537–556.

Bizzo, Nelio. *Darwin no Telhado dos Andes*. São Paulo: Odysseus, 2009.

Bizzo, *Nelio. Ensino de ciências: fácil ou difícil*? São Paulo: Biruta, 2nd ed., 2012.

Bizzo, Nelio. "Understanding and Acceptance of Evolution: Research on Geological Time and Cognition." In Patariya, M. K., M. I. Nogueira. (Orgs.), *Sharing Science*. 1st ed., 79–94. New Delhi: National Council for Science and Technology Communication, 2011.

Bizzo, Nelio. "História da ciência e o ensino da ciência: instrumentos para a prática e a pesquisa escolar." In Valéria A. Arantes (Org.), *Ensino de Ciências*, 13–59. São Paulo: Summus Editorial, 2013.

Bizzo, Nelio, and Adela Molina. "El mito Darwinista en el aula de clase: un análisis de fuentes de información al gran público." *Ciência & Educação (Bauru) 10*, no. 3 (September 2004): 401–416.

Bizzo, Nelio, and Luís Eduardo Bizzo. "Charles Darwin in the Andes." *Journal of Biological Education 40*, no. 2 (June 2006): 68–73.

Bizzo, Nelio, and Charbel Niño El-Hani. "Darwin and Mendel, evolution and genetics." *Journal of Biological Education 43*, no. 3 (September 2009): 108–114.

Bizzo, Nelio, and Jefferson B. Oliveira. "Tempo Geológico nas perspectivas histórica, epistemológica e sócio-cultural: resultados de uma pesquisa transnacional." Proceedings VIII ENPEC: Campinas, R1667-1 (November 2011). Available at: http://www.nutes.ufrj.br/abrapec/viiienpec/resumos/R1667-1.pdf [accessed April 2019].

Bizzo, Nelio, and Jefferson B. Oliveira. "Giambattista Brocchi (1772–1826) e as paleoheteromorfias na alvorada do século XIX". *Filosofia e História da Biologia 7*, no. 2 (July–December 2012): 281–303.

Bowler, Peter. "Darwin on Man in the Origin of Species: A reply to Carl Bajema." *Journal of the History of Biology 22*, no. 3 (September 1989): 497–500.

Clough, Elizabeth E., and Colin Wood-Robinson. "Children's Understanding of Inheritance." *Journal of Biological Education 19*, no. 4 (December 1985): 304–310.

Cooke, Kathy. "Darwin on Man in the Origin of Species: An addendum to the Bajema-Bowler Debate." Journal of the History of Biology 23, no. 3 (September 1990): 517–521.

Darwin Correspondence Project, "Letter no. 275," accessed on 17 March 2019, http://www.darwinproject.ac.uk/DCP-LETT-275.

Darwin, Charles Robert. *The Variation of Animals and Plants under Domestication.* (2nd ed.) London: John Murray, 1875.

Darwin, Charles Robert. *A Origem das Espécies*. Porto: Lello & Irmão, 1946.

Darwin, Charles Robert. *A Origem das Espécies*. São Paulo: Hemus Editora, 1981.

Darwin, Charles Robert. *A Origem das Espécies*. São Paulo: EDIPRO, 2018.

Deadman, John A., and Peter Joseph Kelly. "What do secondary school boys understand about evolution and heredity before they are taught the topics?" *Journal of Biological Education 12*, no. 1 (March 1978): 7–15.

Deadman, John A. "*The structure and development of concepts associated with the topic of evolution in secondary school boys.*" PhD thesis, University of London, 1976.

Dominici, Stefano. "Brocchi's Subapennine Fossil Conchology." *Evolution: Education and Outreach 3*, no. 4 (December 2010): 585–594.

Dominici, Stefano, and Niles Eldredge. (2010). "Brocchi, Darwin, and transmutation: Phylogenetics and paleontology at the dawn of evolutionary biology." *Evolution: Education and Outreach 3*, no. 4 (December 2010): 576–584.

Good, Ronald G., John E. Trowbridge, Sherry S. Demastes, James H. Wandersee, Mark S. Hafner, and Catherine Cummins, (eds.). Proceedings of the 1992 Evolution Education Research Conference. Louisiana State University: Baton Rouge, 1992.

Hokayem, Hayat, and Saouma BouJaoude. "College students' perceptions of the theory of evolution." *Journal of Research in Science Teaching 45*, no. 4 (April 2008): 395–419.

Kargdo, Dennis B., Edward D. Hobbs, and Gaalen L. Erickson. "Children's beliefs about inherited characteristics." *Journal of Biological Education 14*, no. 2 (June 1980): 137–146.

Kesser, Seymour. "Psychological aspects of genetic counseling: VI. A critical review of the literature dealing with education and reproduction." *American Journal of Medical Genetics 34*, no. 3 (November 1989): 340–353.

Krasilchik, Myriam. "Trajetória de uma professora de Biologia." In Carvalho, Anna M. P., Antonio Cachapuz, and Daniel Gil-Perez (eds.), *O ensino das ciências como compromisso científico e social: os caminhos que percorremos*, 197 213. São Paulo, SP: Cortez, 2012.

Lester, Alan, and Dussart, Fae. *Colonization and the Origins of Humanitarian Governance.* Cambridge: Cambridge University Press, 2014.

Macedo-Souza, Lucia I., Fernando Kok, Silvana C. dos Santos, Simone C. Amorim, Alessandra Starling, Agnes Nishimura, Karina Lezirovitz, Angelina M. M. Lino, and Mayana Zatz. "Spastic Paraplegia, Optic Atrophy, and Neuropathy Is Linked to Chromosome 11q13." *Annals of Neurology 57*, no. 5 (May 2005): 730–737.

Martill, David, Gunter Bechly, and Robert F. Loveridge. *The Crato Fossil Beds of Brazil: Window into an ancient world.* Cambridge: Cambridge University Press, 2007.

Oliveira, Jeferson B. "O tempo geológico no ensino fundamental e médio: os estudantes e os livros didáticos." PhD thesis, University of São Paulo, 2006.

Piaget, Jean, and Rolando Garcia. "Psicogénese e História das Ciências." Lisboa: Dom Quixote, 1987.

Rogoff, Barbara, and Jean Lave. "Everyday Cognition." Cambridge, MA: Harvard University Press, 1999.

Santos, Silvana C. dos. "Para geneticistas e educadores: o conhecimento cotidiano sobre herança biológica." PhD thesis, University of São Paulo, 2003.

Santos, Silvana C. dos. "The diversity of everyday ideas about inherited disorders." *Public Understanding of Science 15*, no. 3 (July 2006): 259–275.

Santos, Silvana C. dos, and N. Bizzo. "From 'New Genetics' to everyday knowledge: ideas about how inherited diseases are transmitted in two large Brazilian families." *Science Education 89*, no. 4 (July 2005): 564–576.

Smith, John K., and Louis Heshusius. "Closing down the conversation: The end of the quantitative-qualitative debate among educational inquirers." *Educational Researcher 15*, no. 1 (January 1986): 4–12.

Wertsch, James V. *Vygotsky and the social formation of mind.* Cambridge, MA: Harvard University Press, 1985.

Wertsch, James V. *Voices of collective remembering.* New York: Cambridge University Press, 2002.

Chapter Nine

Comparative Studies of Students' Beliefs and Understandings in Brazil, Italy, and Galapagos Islands

Graciela S. Oliveira, Helenadja Santos Mota, Ana Maria Santos Gouw, and Nelio Bizzo

In recent years, we have observed cultural changes that oscillate between traditional and conservative attitudes toward family values and more flexible and tolerant attitudes about diversity. Such oscillations have permeated and often assumed clear opposition to scientific topics as it is the case of questions involving biological evolution. The view that people present about the origin of the universe and biological diversity on Earth has reached the political sphere in many countries, which has triggered discussions about its teaching in schools. To elucidate such a matter, it becomes necessary to reflect deeply on the several aspects involved in the understandings and beliefs about such topics. This work intends to present some aspects that may help in this discussion, based on research carried out in Brazil, Italy, and the Galapagos Islands in Ecuador.

International studies on cross-cultural similarities and differences can contribute to the understanding of the education systems adopted by the world. New ideas and practices can be identified and improved from comparative studies, as well as the implementation of public policies based on real and successful experiences (Gilford 1993). However, international comparisons must take into account regional peculiarities in order to avoid misunderstandings when exploring and generalizing results out of their proper context. Differences in cultural contexts may affect what, how, and when a given subject is taught, so international investigations need to recognize the effects of cultural differences on student performance (Gilford 1993).

In line with the concerns mentioned above, collaborative work between research groups in Brazil, Italy, and Ecuador was designed after a careful study of previous research, and carried out within the project "Relevance of Science Education" (ROSE), under the coordination of Professor Svein Sjøberg of the University of Oslo. The three countries have different languages, culture, and many other particular aspects; however, the issue of biological evolution was considered highly relevant by the researchers involved in the three settings. As such, it was considered worthwhile to understand how similar public perceptions might be and search for correlations between knowledge and acceptance of biological evolution and science and technology in general.

EVOLUTION EDUCATION

The teaching of biological evolution in different parts of the world occurs in the context of a global curricular change, which has emphasized predominantly greater proximity of the general population to science (Hanley, Bennett, and Ratcliffe 2014). The theory of evolution is widely recognized for providing the ideal illustration of several aspects of the nature of science; the role of evidence to support scientific theories; the process of development of theory in the biological sciences; and the relationship between sociocultural knowledge and scientific knowledge (Hanley, Bennett, and Ratcliffe 2014).

Bizzo and El-Hani (2009) identify several additional reasons that justify the presence of evolutionary theory in K–12 education: 1) it is central and fundamental in the study of all life sciences and socio-scientific issues, such as genetic engineering, resistance to antibiotics, conservation, agriculture, and other subjects; 2) it fosters evolutionary thinking that places the human being as part of the world of living beings, not from outside and above it, having significant consequences for the way the individual relates to the rest of the organisms and deals with environmental issues and with animals rights; 3) it favors discussions about the nature of knowledge and, more specifically, the nature of science; 4) it enables a more coherent and profound understanding of human health and diseases; and, 5) the origin of biology as a science, as an integrated field of knowledge about living systems, is related to evolutionary thinking.

Anderson (2007) comments that the relation of the young students with the theory of biological evolution happens in three multifaceted and complex, interrelated contexts: the social, the intellectual, and the pedagogical. As to the pedagogical aspect, when biological evolution is included in the curriculum of science classes, most of the curricular guidelines focus on the conceptual content. It is even rarer for educators to pay any attention to philosophical ideas linked to theory or how the study of biological evolution relates to students' daily lives and worldviews.

In a classroom, students' values and beliefs reflect those of the general population, which are very varied. Therefore, it is useful to consider general population patterns in scientific research. Anderson (2007) comments that in the United States, for example, classrooms reflect a skeptical population regarding biological evolution, with incipient knowledge about basic concepts of theory, and there is a strong tendency to view theory less credible than other scientific concepts (Alters and Alters 2001).

Concerning the values and beliefs related to the origin of life, Anderson (2007) describes how elementary students' previous conceptions are connected to religious ideas, a natural result of their relations with the social environment. In the case of young North Americans, the author comments that religious students probably have a connection with the religion greater than that recognized by the school. Thus, ignoring the cultural perspectives of students and moving forward with the teaching of evolutionary theory prevents the learning process from actually taking place, since the skepticism of some social groups is related to their religious perspective.

On the values attributed to the theory of biological evolution and religion in Brazil, we find research conducted by the group coordinated by Nelio Bizzo since 2007. The first initiatives resulted in the data analyzed by Oliveira (2009) and Oliveira and Bizzo (2011), who verified the acceptance/rejection of the theory of evolution by eighth-grade students from public schools located in two very distinct settings, Tangará da Serra (MT) and São Caetano do Sul (SP). The first is a predominantly rural area in central Brazil, whereas the second is a highly industrial city in the southeast. The aim was to seek the possible relations between the attitude of informants about evolutionary theory and proximity to science and religion.

Shortly after, the same instrument was used in a study carried out in Brazil and Italy in the context of the ROSE project. The data collection instrument of the project, a questionnaire composed of 245 items, aimed at investigating the opinions of fifteen-year-old students about their experiences in science classes and their relationship with science in general and future career possibilities. In Brazil, a section on biological evolution was added to the ROSE instrument as a national section (Santos Gouw 2013).

In Brazil and Italy, research findings also indicated that young people viewed science and technology as being of high interest and social importance, although they did not hold the same views of their benefits or showed any interest in experiencing scientific and technological processes directly or in pursuing a scientific career. In 2013, as a follow up to the work already carried out in both countries, a data collection instrument called *Barômetro* was developed through a reduction of the ROSE questionnaire, informed by data obtained from the two countries regarding the concerns, interests, and perceptions of young people about science, their schools, science classes and scientific subjects. Furthermore, the section on biological evolution was reviewed in both languages, specifically targeting understanding and acceptance of biological evolution (Bizzo and Pellegrini 2013).

The *Barômetro* instrument was used with a sample of national representativeness in both Brazil (N = 2404) and Italy (N = 3503) in 2014. The *Barômetro* instrument was later implemented in the Galapagos Islands (N = 385), Ecuador, in the year 2015.

The results found in the three countries suggest a need for deeper reflections about the educational realities of schools and a better understanding of their relations with science and scientific subjects. As a step in this direction, the current chapter presents the main results of this exploratory work about the relationship between youth and the theory of biological evolution in Brazil, Italy, and Galapagos Islands (Ecuador). More specifically, we present the main similarities and differences identified in the answers of the students sampled across the three countries.

METHODOLOGY

The research presented here is primarily quantitative and reports the results of the application of the *Barômetro* questionnaire between 2014 and 2015 to samples of national representativeness in Italy and Brazil and of local representativeness in the Galapagos Islands. In the three localities, the instrument was applied to students with a mean age of fifteen years old.

In Brazil, a nationwide sample of students enrolled in the first year of high school in public and private schools was assembled. Sample design informed the selection process stratified by conglomerates in two stages: school and class. Schools were selected by drawing from the PISA sample, reaching 100 schools in the five regions of the country. We received questionnaires answered from 78 schools, located in 72 municipalities from all five regions of Brazil. The total sample of students was 2,404 (55.1 percent girls). As for the age group of young people who participated in the research, there was a higher frequency of students with fifteen years of age (47.4 percent).

The data collection strategy adopted in Brazil was sending questionnaires by mail after previous contacts with the schools by phone, followed by the formal submission of the written research project by e-mail. Telephone contacts took place between March and May 2014, with the 100 schools drawn; however, in the initial contacts, some declined to participate and, therefore, a reserve list of drawable schools (N = 20) was used in order to guarantee the sending of the questionnaires to 100 schools, as stipulated in the sample plan. The application of the instrument was carried out by a collaborating local teacher, from May to July 2014, in a class of first-year high schoolers. The application of the questionnaires took place in classrooms, and most of the teachers stated that thirty minutes were enough to present the objectives, clarifications on filling in, distributing and obtaining the students' answers.

In Italy, target students were in the second year of secondary education, enrolled in T-level schools ("Istituti Professionali") or propaedeutic high schools (*"Licei")* belonging to one of four types: classical, scientific, linguistic, and artistic. The questionnaires were applied to students from the Northwest, Northeast, Center, and South, and Islands. According to the sampling plan, the Italian research team also used stratified sampling and obtained the representation of the universe of Italian students from two predetermined variables: 1) the number of schools by geographical location; and, 2) the number of students enrolled in the second year of secondary education.

The schools that participated in the research were selected by drawing, and as a result 100 schools were invited to participate in the project. Almost all of them (99) returned the questionnaires. Each school counted on the teachers' contribution to the application of the questionnaires, which took place between March and May 2014. The sample reached 3,503 students (52 percent girls). The majority of the students who answered the questionnaire in Italy were between fifteen and sixteen years old (77.3 percent) at the time of data collection.

The Galapagos Islands have a population of about 25,000 people and about 590 students aged fifteen were starting high school (2015). Participating in this research were students enrolled in the 1st year of the General Unified Bachelor in the Educational Units of Galapagos. A total of 23 institutions were registered in the archipelago, of which 9 offered the first year of bachelor degree. These institutions are located in the islands of Santa Cruz (5), Isabela (1) and San Cristóbal (3), with a total of 518 students. The research was carried out, in May, 2015, in 6 institutions, and a total of 385 students participated in the sample, which is equivalent to 74.3 percent of the universe of the study. Of the total sample, 54.5 percent were boys (N = 210).

The *Barômetro* questionnaire was structured based on a reduction of ROSE and presented items in the form of four-point Likert type scale, true or false, and multiple-choice questions. The present work deals with the sections that examined possible relationships between students of religious background and the theory of biological evolution. In the religion section, we sought to ascertain how many students considered to be followers of a religion. Next, students were asked to mark the religious congregation to which they belonged (multiple-choice format).

In the section on views and conceptions on the theory of biological evolution, there were statements recognized by the scientific community that framed some key ideas of the theory of biological evolution, such as common ancestry, origin and age of the Earth, gradual changes of living organisms, as well as human origins. After many years working on the theme (see previous chapter), it was clear to the research team that some key points had to be addressed in order to assess students' understanding, including: 1) the Earth is very old; 2) species evolve from common ancestors; 3) evolution encompasses all species, humans included. Items design followed a previous survey, where Brazilian students answered similar items (Mota 2013).

The data were analyzed using the *Software Statistical Package for Social Science* (SPSS)—Statistical Package for Social Sciences—version 18.0. The data were initially presented in a descriptive way, represented by absolute numbers, relative and mean/standard deviation for each variable referring to the relation of the young with the theory of biological evolution.

EVOLUTION VIEWS AND UNDERSTANDINGS ACROSS COUNTRIES

In all three countries, students' knowledge of concepts related to the theory of biological evolution was evaluated, raising topics considered valid by students. The young people responded by considering the claims in the questions as true, false, or declining to answer. Additionally, the views of participants were studied by religious groups to characterize the influences of social and cultural background on their responses.

Brazilian, Italian, and Ecuadorian students answered questions related to evolutionary knowledge about the age of the Earth, fossil records, change of living beings, attribution of the concept of biological evolution to nature, common ancestry, and the history of mankind on Earth. Besides, a statement about the coexistence of humans and dinosaurs was included in the questionnaire to verify how deep their knowledge on geological time was. The results of Brazilian students were:

- *The formation of our planet occurred some 4.5 billion years ago*: 9.5 percent False, 46.4 percent True, (44 percent declined);
- *Fossils are evidence of living things that lived in the past*: 6.6 percent False, 12.3 percent True, (81 percent declined);
- *The present species of animals and plants originated from other species of the past*: 12.6 percent False, 63.4 percent True, (23.8 percent declined);
- *Biological evolution occurs in both plants and animals*: 11.8 percent False, 66.1 percent True, (22 percent declined);
- *Human species originated from another primate species*: 27.8 percent False, 41.1 percent True, (31 percent declined);
- *Human species have inhabited planet Earth in the last 100,000 years*: 14.8 percent False, 29.3 percent True, (55.8 percent declined);

- *Different organisms may have a common ancestor*: 18.7 percent False, 29.8 percent True, (51.3 percent declined);
- *The first humans were prey to carnivorous dinosaurs*: 37.4 percent False, 22.2 percent True, (40.3 percent declined).

In general, low percentages of Brazilian students considered statements to be correct, and they were less responsive in the questions about the meaning of fossils, age of the Earth, origin of mankind, how long humans have been on Earth, and common ancestry. It is likely that they had limited knowledge about the meaning of the word in the sense that they may have never seen a fossil. About the coexistence of humans and dinosaurs, there was a higher percentage of students declining to answer (40.3 percent) or choosing "True" (22.2 percent); this means that two-thirds of the students did not have a reasonable glimpse of past events on the history of life.

A small percentage, 29.3 percent, agreed that humans had inhabited Earth one hundred thousand years ago, but 18.7 percent considered the statement false, and 51.3 percent declined to respond. This result supports the idea that some two-thirds of the students do not have a proper view of past events on the history of mankind. Students seemed to be uncertain about the claims on human evolution, as 41.1 percent considered humans and primates to have common ancestors.

The results found among Italians students in the same questions were strikingly different:

- *The formation of our planet occurred some 4.5 billion years ago*: 10.5 percent False, 71.9 percent True, (17,5 percent declined);
- *Fossils are evidence of living things that lived in the past*: 2.1 percent False, 94.6 percent True, (3.2 percent declined);
- *The present species of animals and plants originated from other species of the past*: 4.9 percent False, 86.2 percent True, (8.8 percent declined);
- *Biological evolution occurs in both plants and animals*: 7.5 percent False, 71.7 percent True, (20.7 percent declined,);
- *Human species originated from another primate species*: 5.9 percent False, 84.2 percent True, (9.7 percent declined);
- *Human species have inhabited planet Earth in the last 100,000 years*: 29.2 percent False, 19.7 percent True (51 percent declined);
- *Different organisms may have a common ancestor*: 10.9 percent False, 61.6 percent True, (27.4 percent declined);
- *The first humans were prey to carnivorous dinosaurs*: 61.6 percent False, 12.4 percent True, (25.9 percent declined).

The Italian students had a higher percentage of right answers about the topics on the theory of biological evolution, with a greater tendency to select the "true" option, indicating that there is no doubt about the age of the Earth (71.9 percent); that fossil records are evidence of organisms that lived in the past (94.6 percent); that both plants and animals evolved (71.7 percent); common ancestry among species (61.6 percent); as well as to the ancestry of humans from another species of primate (84.2 percent). However, when faced with the statement "The human species inhabited the Earth about one hundred thousand years ago," they seemed confused. There was a much higher frequency in the neutral alternative (51 percent) followed by the "False" option (29.2 percent). This may be a consequence of some insecurity regarding the exact number of years. For instance, students who had in mind a different but similar number (say, two hundred thousand years), may have restrained from considering the statement as true. The coexistence of humans and dinosaurs was discredited by the majority of students (61.6 percent) showing a much higher frequency of proper views of the past ages.

The results found among students of the Galapagos Islands were:

- *The formation of our planet occurred some 4.5 billion years ago*: 9.6 percent False, 40.1 percent True, (50.3 percent declined);
- *Fossils are evidence of living things that lived in the past*: 3.2 percent False, 87.9 percent True, (8.9 percent declined);
- *The present species of animals and plants originated from other species of the past*: 7.4 percent False, 75.7 percent True, (8.9 percent declined);
- *Biological evolution occurs in both plants and animals*: 3.9 percent False, 88.4 percent True, (7.7 percent declined,);
- *Human species originated from another primate species*: 24.5 percent False, 48.2 percent True, (27.4 percent declined);
- *Human species have inhabited planet Earth in the last 100,000 years*: 9.5 percent False, 30.4 percent True, (60.1 percent declined);
- *Different organisms may have a common ancestor*: 7.4 percent False, 52.3 percent True, (40.3 percent declined);
- *The first humans were prey to carnivorous dinosaurs*: 40.1 percent False, 20.6 percent True, (39.3 percent declined).

A high percentage of students from the Galapagos Islands had a good score on some items, especially when recognizing fossils as evidence of living beings (87.9 percent) and admitting that plants also evolve (88.4 percent). Soria and Adrián (2016) puts forward the idea that in the island of Santa Cruz there are displays in many places showing endemic plants, like the giant daisy trees from the Asteraceae family, *Scalesia* sp, with at least fifteen native shrubs

and tree species and six subspecies with very particular life cycles. Schools have been targeted in conservation programs, and a zone of highlands of some of those islands is referred to as the "Scalesia zone." The percentages of agreement with evolutionary origin of humans is surprisingly high (48.2 percent), especially when compared with the silence about geological time, as many students declined to answer item on the age of Earth (50.3 percent), and the age of humans on Earth (60.1 percent).

From the answers of the Brazilian and Ecuadorian students, it was possible to identify three distinct categories: 1) *concepts recognized as correct*—content items whose truthfulness was accepted by the majority of young people such as species change and attribution of the concept of evolution to plants and animals; 2) *concepts that seem confusing*—particularly frequent among Brazilians, including a large number of students who declined to answer questions such as the meaning of fossils, age of humans on the planet, formation of Earth; and, 3) *unclear concepts*—items which are not recognized as correct by a reasonable amount of students, such as possible coexistence of humans and dinosaurs, reveal that geological time is hardly understood. The figure is much different considering Italian students, who show a more accurate and consistent knowledge about the key concepts related to the evolutionary theory in the first grade of high school.

These results are consistent to those found by Mota (2013), who stressed that young Brazilian students seemed willing to agree with statements about fossil records, but they were uncertain about items that address common ancestry and tended to reject those related to human evolution. However, the author considers that, in general, these students hold somewhat positive attitudes toward evolutionary theory, considering the high percentage of participants who declared to belong to religions which tend to deny the evolutionary perspective (see ahead).

Nonetheless, despite the positive attitudes of young people regarding evolutionary topics, it is evident that knowledge of Brazilian and Ecuadorian students is relatively low. While they consider the perspective of biological change of organisms as valid and understand it as a natural phenomenon, more complex biological concepts seem to be either confusing or unclear to them, such as common ancestry and geological time. In contrast, the results from Italy show that there is generally a more in-depth understanding of evolutionary issues and that students recognize the validity of different aspects of evolutionary theory as well as human evolution.

Although there are sociocultural differences that markedly limit more in-depth comparisons of the realities of Brazil, Italy, and the Galapagos Islands, the responses of youngsters across the three countries point to some apparent trends. It was clear that Italian students are better informed about evolution than Brazilian students. This difference is possibly due to some extent to differences in the educational systems. In the Italian case, science education includes discussions about human ancestry since the first years of primary school (D'Ambrosio 2018). In Brazil, the situation may be strikingly different, as young Brazilians may have had little or no contact with the idea of descent with modification alongside all K–12 school years.

In Galapagos, students have access to many references to Darwin, and the national curriculum ("Nuevo Currículo de la Educación Básica") includes a brief introduction to the theory of evolution. However, the textbook used in the Galapagos Islands (Valverde 2013) provided a succinct summary of the human evolution, a short phase only ("There is evidence of a common ancestor between simians and modern humans, as a consequence of a common and parallel evolutionary history" 26). Creationism is taught to students as another "theory" alongside biological evolution in the same grade 9 (Soria, 2016). Besides, there is strong religious ubiquity in the islands; the largest school in the island of Santa Cruz, which took part in the present research, has its walls artistically painted with phrases of the biblical book of Genesis describing the creation. Although Ecuadorian students find references to Charles Darwin in a number of places in Galapagos, mainly in Santa Cruz and San Cristóbal islands, some of them referred to him as a "naturista" ("naturist"), a sun worshipper, misinterpreting the word "naturalista" ("naturalist") (Soria and Adrián, 2015, personal communication to Nelio Bizzo).

Some students may find it challenging to accept the evolutionary theory as valid due to the lack of sufficient familiarity with or understanding of the evidence supporting the theory (Allmon 2011). In the present study, students' responses to questions about the age of Earth or the meaning of fossils provide support for this proposition.

There are also sociocultural factors, which can interfere with the opinions of young people, such as religion. Religious affiliation constituted an essential topic for the students surveyed. In the next section, the responses of young people by religious groups are presented.

BIOLOGICAL EVOLUTION AND RELIGION

Religion continues to influence various aspects of society, such as economic, political, social, and cultural spheres (Davie 2010). It is no different in how young people relate to scientific explanations for the origin and evolution of life on Earth. In this sense, it becomes relevant to understand the relationship of students with religion in diverse cultural contexts and how these relations influence their knowledge and acceptance of evolutionary theory.

In the religion section, we sought to verify the number of students who declared belonging to any religion (90.4 percent of girls and 85.3 percent of boys). Next, these students had to mark the religious congregation to which they belong to (multiple choice format). We found the following distribution in the Brazilian student sample regarding religious affiliation: Catholics (girls 55.8 percent; boys 56.5 percent); Evangelicals of the mission (girls 10.6 percent; boys 9.7 percent); Pentecostal evangelicals (Girls 20.5 percent; Boys 21.7 percent); and, other religion (girls 13.2 percent; boys 12.1 percent).

In the questionnaire applied in Italy, the organization of this section was different as students had to choose among the following items: *1) yes, a Catholic Christian religion* (Girls 68.9 percent; Boys 64.9 percent)*; 2) yes, a non-Catholic Christian religion (Orthodox, Lutheran, Valdense . . .)* (girls 3.4 percent; boys 2.6 percent)*; 3) yes, other religions (Hebrew, Buddhist, Islamic)* (girls 3.7 percent; boys 3.6 percent)*; 4) yes, a New Age philosophy* (girls 3.8 percent; boys 4.1 percent)*; 5) No, I do not believe in any religion or transcendental philosophy* (girls 19.9 percent; boys 24 percent). Italy is a predominantly Catholic country with a recognized diversity of religious expressions, but not as marked as in Brazil.

In the Galapagos Islands, after answering the item about belonging to a religion (84 percent), these students declared their religious affiliation by choosing among the following items: Catholic (66.3 percent), Evangelical (14.6 percent), Christian (4.6 percent), Witnesses of Jehovah (4.3 percent), Adventist (4 percent), and Mormon (3.4 percent). Table 9.1 shows the variety of responses of Brazilians by their religion.

In the Brazilian sample, there was a significant difference among religious groups regarding statements about the age of the Earth, fossil records, the attribution of biological evolution to nature. A more extreme difference was seen in the item that states that humans descend from another primate species (see numbers in Table 9.1), in which agreement with the statement correlates highly with religion (or no religion).

In Italy, students belonging to other religions, including Hebrews, Buddhists, and Islamists chose the "true" option less frequently in response to the item about the fossils as records of living beings who lived in the past (86.5 percent). Non-Catholic Christians (Orthodox, Lutheran, and Waldensian) students also showed a lower percentage when asked whether the following statement was true: "The present species of animals and plants originated from other species of the past" (non-Catholic Christians: 79.2 percent, other religions: 79.4 percent); "Biological evolution occurs in both plants and animals" (non-Catholic Christians: 68.3 percent, other religions: 68 percent); "The human species descent from another primate species" (non-Catholic Christians: 66 percent, other religions: 66.1 percent); "Different organisms

Table 9.1. Distribution of the sample in Brazil regarding knowledge about biological evolution, by religious group.

		Catholics	*Evangelicals of mission*	*Pentecostal Evangelicals*	*Other Religion*	*Without religion*	*p-value*
				%			
I-The formation of our planet occurred some 4.5 billion years ago	T	48,4	39,6	40,5	47,6	49,4	0,017
	F	7,7	12,1	11,8	9,7	0,7	
	D/K	43,9	48,3	47,6	42,7	43,5	
II-Fossils are evidence of living things that lived in the past	T	84	75,8	78	77	86,7	0,017
	F	5,5	7,7	7,9	8,1	5,5	
	D/K	10,5	16,4	14,1	14,9	7,7	
III-The present species of animals and plants originated from other species of the past	T	65,9	58,7	59,5	63	67,4	0,012
	F	10,2	17	16,1	14,6	9,3	
	D/K	24	24,3	24,5	22,4	23,3	
IV-Biological evolution occurs in both plants and animals	T	67,3	64,5	66,7	65	65,8	0,732
	F	10,6	14,3	12,7	12,8	11,2	
	D/K	22,1	21,2	20,6	22,2	23	
V-Human species descends from another primate species	T	47,6	25,7	31,5	35,5	47,4	0,000
	F	21,7	48,1	37,2	31,8	20,7	
	D/K	30,7	26,2	31,3	32,7	31,9	
VI-The human species has inhabited planet Earth in the last 100,000 years	T	29	31,9	29,3	27,3	29,7	0,927
	F	15,1	15,7	14,6	17,1	12,6	
	D/K	55,9	52,5	56,1	55,5	57,6	
VII-Different organisms may have a common ancestor	T	30,8	26,2	27,3	25,3	38,6	0,497
	F	18,2	18,8	20,7	20,4	15,7	
	D/K	51	55	52,1	54,3	45,7	
VIII-The first humans were prey to carnivorous dinosaurs	T	23,5	18,7	22,8	19,7	20,9	0,575
	F	35,4	39,4	38,3	39,3	38,1	
	D/K	41,1	41,9	39	41	41	

may have a common ancestor" (non-Catholic Christians: 55 percent, other religions: 48.8 percent). However, it is essential to consider the sample sizes as the numbers do not allow a consistent comparison with Christian Catholic students.

In the item *The human species has inhabited the Earth about 100,000 years ago,* there was a difference in the responses of non-Catholic Christian Italian students and non-believers. As for the coexistence of humans and dinosaurs, young Christian non-Catholics and other religions had higher percentages when considering the correct statement, with 15.8 percent and 19.4 percent, respectively.

Significant differences between religious denominations were identified in seven items for Italians, and in four items for Brazilians (Table 9.1, items I, II, III and V; for details, see Oliveira, Bizzo and Pellegrini 2016). According to Soria (2016), the difference in the responses of students by religious denomination was lower in the Galapagos Islands, because significant p-value was identified only in the items IV ("Biological evolution occurs in both plants and animals"; p-value 0.001) and V ("The human species originated from another primate species"; p-value 0.001).

In Italy, differences in the responses of students can be identified by religious denominations, whereas in the Brazilian data this appears to a lesser extent. In general, it can be seen that in the three countries some religious groups showed a greater tendency to identify topics in the theory of biological evolution as true. However, in Italy, the differences appear in higher proportion.

A possible inference would be that students belonging to all religious denominations in Italy have a higher level of information and therefore differences among them tend to be more explicit, whereas in the other two countries the lower level of information leads to unavoidable leveling of different religions regarding issues about geological time and evolution.

Religious ideas may play unique roles for the students surveyed since for some Brazilian students, spiritual knowledge seems to fill in possible knowledge gaps about evolutionary theory, whereas for Italians it may represent a complementary positioning in the face of science, thus forming a different way of describing and explaining the world.

The data results from Brazil suggest that the more significant knowledge gaps may play a far more critical role than religion in explaining their opinions about evolution. The data indicate that Brazilian students tend to decline to answer questions, including non-believers. The high frequency of those who refuse to answer could show that students recognize that they do not hold enough explicit knowledge to stand a positive opinion when facing an objective position. In Italy, on the contrary, students of all religions and non-believers feel more confident to answer either positively or negatively.

In the case of Brazil, it is likely that conceptual gaps or misconceptions regarding evolutionary theory have an overwhelming influence on the results, compared with details of the religious doctrine of different religions. Thus, general aspects of culture, and not the particularities of different religions would have explanatory power over the overall picture.

Last, the direct contact with the Galapagos Islands elements, such as finches, mockingbirds, marine iguanas, and tortoises may not be enough evidence to change students' ideas on evolution. Contrary to normal expectations, local students from different religions had rather similar answers, which may reflect a similar situation, that is, local culture plays a critical role to explain students' answers. In the case of the Witnesses of Jehovah, despite accepting geological time, most students refused to answer the questions about the age of the planet, but they roughly tend to agree with evolutionary biology (Soria 2016)

CONCLUSIONS

The present study has produced evidence that generally points out to different sociocultural influences in the opinions of young people regarding evolution. Some students know some evolutionary topics well and respond accordingly to prompts involving topics such as the origin of humans. However, other students who lack critical knowledge on the subject may refrain from answering objectively to a prompt involving a topic such as the age of Earth. Their position does not reflect neither their general religious beliefs nor particularities of the theological doctrine they follow, but the overall influence of their sociocultural environment. In the Galapagos Islands, there is an overall tendency to praise Darwin highly, as the islands' economy heavily depends on the so-called "scientific tourism." This is different in Brazil, where Pentecostal religions have increased and have a radical discourse against evolution.

A similar study was carried out with students of the same age in two schools of the metropolitan area of Rio de Janeiro. One of them is in Nilopolis, a town with a very significant presence of Pentecostal churches. In the surroundings of the school, there were some fifteen Pentecostal churches. That school had 43.7 percent of students belonging to this religious group. The sample included data from students of another school, where the presence of Pentecostal students was lower (15 percent). Results suggest that Pentecostal religions tend to be the most critical isolated factor to predict belief in the biblical narrative, opposing the evolutionary perspective (Teixeira 2019).

Our data, from the nationwide sample in Brazil carried out in the same period and with the same age level, did not find significant differences between the two groups of evangelicals (table 9.1).

Data from nationwide samples in Brazil taken in 2010 and 2014 and the Galapagos sample show that Evangelical students do not reject evolutionary perspective as much as expected, taking into account the particularities of the different Evangelic doctrines (Bizzo, Gouw, and Mota, 2013; Oliveira, Bizzo and Pellegrini 2016; Bizzo et al. 2018). Another study carried out in the northern region of Brazil found similar results, as students from different religions, including Evangelicals, tend to accept the evolutionary perspective, except when considering the origin of humans (Almeida 2011). These results can be seen as a result of the local culture, and not directly from the details of the doctrine of the Christian religious denominations.

Human diversity brings to the classroom diverse cultures, and this seems particularly critical for science education since not all students can succeed in science (Cobern and Loving 2008). It is not always evident to students when a problem requires a scientific solution (Cobern and Loving 2001). Thus, the results presented here reinforce that science classes are not exempt from social and cultural implications and that the beliefs of young people cannot be just ignored in the classroom. Diversity is present and confirms that different discourses are interacting and possibly some students need more support than others to engage in the scientific perspective.

Everyday life requires a diversified repertoire of conceptual knowledge. Thus, schools and educators have to develop curricula that value knowledge in its various forms and sources (Cobern and Loving 2001). Plurality provides opportunities for several discourses to coexist and interact, as the multiple species coexist and interact (Hokayem and Boujaoude 2008). The new suggestion of the "pedagogy of difference" may help teachers enhance students' epistemic repertoire through recognition of the multiple approaches between science and religion (Owens et al. 2018).

People have many ways of constructing reality, and each type of knowledge can have privileges only within its limits and its purpose (Cobern and Loving 2008). In this sense, science education can represent an opportunity to cross borders between the everyday world and the school environment, allowing students to maintain respect for their own culture, but activating a cross between cultures, that is, between the knowledge originated of the student's home environment and scientific knowledge (Aikenhead 2001). The challenge is to give voice to the students in the process of teaching and learning, and be sensitive to the different views and, mainly, the different outlooks on the world.

Finally, we believe that in-depth knowledge of a controversial topic such as biological evolution does not hinder young people from reassuring their religious beliefs as science is not a threat to any religious segment. In this sense, knowledge is liberating, both for the spiritual reflection and for the understanding of the reality that surrounds us.

DISCLAIMER

Editorial constraints limited the number of tables in this chapter to one only.

ACKNOWLEDGMENTS

Several people and institutions contributed to the funding and implementation of the several pieces of research reported here. We want to thank Svein Sjøberg, Giuseppe Pellegrini, Nicolas Cuvi, Adrián Soria, Jaqueline Pinafo, Ana Paula Pietri, and Tatiana Tavares da Silva. Xavier Fazio and the editors of this book, Alan Oliveira and Kristin Cook provided useful criticism and suggestions. Funding agencies include CNPq (proc. 476205/2013-1, 503896/2010-1, 481634/2009-6), CAPES/MEC-PAEP (proc. 6726/2012-76), CAPES/Observatório da Educação (Edital no 38/2010-Obeduc), FAPESP (proc. 2010/10320-4; 2016/05843-4), EDEVO/Darwin-USP (Pró-Reitoria de Pesquisa), Faculdade de Educação da USP, FLACSO/Ecuador (Quito), and Observa Science in Society (Italy).

REFERENCES

Aikenhead, Glen S. "Students' ease in crossing cultural borders in school science." *Science & Education* 85, no. 2 (February 2001): 180–188.

Almeida, David F. "Concepções de alunos do ensino médio sobre a origem das espécies" [High School students' conceptions about the origin of species]. *Ciência & Educação* 18, no. 1 (2011): 143–154.

Allmon, Warren D. "Why don't people think evolution is true? Implications for teaching, in and out of the classroom." *Evolution: Education and Outreach* 4, no. 4 (December 2011): 648–665.

Alters, Brian J., and Sandra M. Alters. *Defending evolution in the classroom: A guide to the creation/evolution controversy*. Canada: Jones and Bartlett Publishers, 2001.

Anderson, R. D. "Teaching the theory of evolution in social, intellectual, and pedagogical context." *Science & Education* 91, no. 4 (June 2007): 664–677.

Bizzo, Nelio, and Charbel N. EL-Hani. "Darwin and Mendel: Evolution and genetics." *Journal of Biological Education* 43, no. 3 (June 2009): 108–114.

Bizzo, Nelio, and Giuseppe Pellegrini. *Os jovens e a ciência*. Curitiba: CRV, 2013.

Bizzo, Nelio, Giuseppe Pellegrini, Nicolas Cuvi, Graciela S. Oliveira, and Adrián Soria. "Opiniones de los estudiantes sobre evolución biológica: muestras nacionales de Italia y Brasil, y comparación com Galápagos" [Students's opinions abou biological Evolution: nation-wide samples from Brazil and Italy and comparison with Galapágos]. In Vallego, Gustavo, Marisa Miranda, Rosaura R. Guitiérrez, and Migel A. Puig-Samper (ords), *Darwin y el Darwinismo desde el Sur del Sur*, 419–29. Madrid: Doce Calles, 2018.

Bizzo, Nelio, Ana Maria Santos Gouw, and Helenadja Mota. "Evolução e religião: o que pensam os jovens estudantes brasileiros." *Ciência Hoje* 50, no. 300 (February 2013): 26–31.

Cobern, William W., and Cathleen C. Loving. "Defining 'science' in a multicultural world: Implications for science education." *Science & Education* 85, no. 1 (January 2001): 50–67.

Cobern, William W., and Cathleen C. Loving. "An essay for educators: Epistemological realism really is common sense." *Science & Education* 17, no. 4 (April 2008): 425–447.

D'Ambrósio, Marcela. "*O Ensino de Evolução na escola básica do Brasil e da Itália: uma análise comparativa de livros didáticos*." Dissertation, University of Campinas, 2018.

Davie, Grace. "Resacralization." In Bryan S. Turner (ed.), *The new Blackwell companion to the sociology of religion*, 160–177. Malden, MA: Blackwell Publishing, 2010. https: file:///C:/Users/home/Downloads/54117b840cf29e4a232965e7.pdf.

Gilford, D. M. A. *Collaborative Agenda for Improving International Comparative*. Washington, DC: National Academies of Sciences, 1993. http://files.eric.ed.gov/fulltext/ED392806.pdf.

Hanley, Pam, Judith Bennett, and Mary Ratcliffe. "The inter-relationship of science and religion: A typology of engagement." *International Journal of Science Education 36*, no. 7 (2014): 1210–1229.

Hokayem, Hayat, and Saouma BouJaoude. "College students' perceptions of the theory of evolution." *Journal of Research in Science Teaching* 45, no. 4 (March 2008): 395–419.

Mota, Helenadja. 2013. "*Evolução Biológica* e Religião: Atitudes de jovens estudantes brasileiros." PhD thesis, University of São Paulo, 2013.

Oliveira, Graciela S. 2009. *Aceitação/rejeição da Evolução Biológica: atitudes de alunos da Educação Básica*. Thesis. Universidade de São Paulo, São Paulo.

Oliveira, Graciela S., and Nelio Bizzo. "Aceitação da evolução biológica: atitudes de estudantes do ensino médio de duas regiões brasileiras." *Revista Brasileira de Pesquisa em Educação em Ciências* 11, no. 1 (November 2011): 57–79.

Oliveira, Graciela. S., Nelio Bizzo, and Giuseppe Pellegrini. "Evolução biológica e os estudantes: um estudo comparativo Brasil e Itália." *Ciência & Educação* 22, no. 3 (July 2016): 689–705.

Owens, David S., Rachel S. A. Pear, Hanan A. Alexander, Michael Reiss, and Tali Tal. "Scientific and religious perspectives on evolution in the curriculum: An approach based on pedagogy of difference." *Research in Science Education* 48 (Deember 2018): 1171–1186.

Rose. *ROSE—International Partners*. 2012. Accessed October 2018 at http://roseproject.no/?page_id=16.

Santos Gouw, Ana Maria. "*As opiniões, interesses e atitudes* dos jovens brasileiros frente à ciência: uma avaliação em âmbito nacional." PhD thesis, University of São Paulo, 2013.

Soria, Adrián. 2015. Personal Communication.

Soria Robalino, Adrián Fidel. "*Comprensión de la evolución biológica* entre estudiantes en Galápagos." Dissertation, Latin American Faculty of Social Sciences of Quito, 2016.

Teixeira, Pedro. "Acceptance of the theory of evolution by high school students in Rio de Janeiro, Brazil: Scientific aspects of evolution and the biblical narrative." *International Journal of Science Education* 41, no. 4 (2019): 546–566.

Valverde Espinosa, Lucila, E. Ciencias Naturales 9. Quito: Ministerio de la Educacion del Ecuador, 2013.

Chapter Ten

The Perils of the Evolution-As-Progress Metaphor

Challenging Ideals of Naturalness, Normalcy, and Adequacy in Brazilian and Canadian Science Education

Giuliano Reis, Adam Oliver Brown,
Delano Silva, and Ana Julia Pedreira

Maybe I was meant to be a fish. Maybe nature just corrected her error.

—*The Incredible Mr. Limpet* (1964)

INTRODUCTION: CAN CHANGE BE *GOOD*?

In the 1964 movie *The Incredible Mr. Limpet*, Henry Limpet (played by the hilarious Don Knotts) is a man who inexplicably turns into a tilefish-like talking fish and goes to help the US Navy to locate and destroy Nazi submarines. As whimsical as the plot may sound, the fact is that Mr. Limpet feels fulfilled in his new form, so much so that he dislikes the possibility of changing back to his human condition. He believes that nature had a plan designed for him, but *she* had made a mistake by letting him be a *Homo sapiens* in the first place.

Implicit in Mr. Limpet's belief that he was ever intended to swim underwater (and talk!), there are two main assumptions that relate to science, specifically to the notion of change. First, living organisms are *meant to be* whatever they are. In other words, there is some grandiose project created for each individual on this planet by some female non-human entity—what he calls nature—that is beyond anyone's comprehension. (This *teleological* orientation to understanding one's purpose in life can actually be extended to anything [Woodfield 1976]). Second, nature makes mistakes that can—if not must—be corrected as they deviate whatever was brought into existence from its original intent. This chapter is primarily concerned with these inaccurate presuppositions, not in the least because they were in some comedy film script, but because they exist in the real world and mediate how evolutionary science is taught to/perceived by younger generations in schools.

Hollywood pictures aside—and there are many examples of how popular media can play an influential role in the public understanding of science; for example, through science fiction (Trushell 2004)—it is unlikely that any of our readers (wherever they might be) have not heard of Charles Darwin. It is equally difficult to conceive that you may have never been exposed to any elements of his theory of evolution. Even if only serendipitously, Darwin's most notorious work is manifested in several occasions: from the zany Darwin Awards, which commemorates "those who improve our gene pool by removing themselves from it in the most spectacular way possible" (https://darwinawards.com/), to the March of Progress/The Road to *Homo sapiens*, perhaps one of the most recognizable (although unintentionally inaccurate) images in the history of scientific illustration (Conniff and Giller 2015). Indeed, Darwin's *On the Origin of Species* (1859) is considered second only to the Bible in terms of the magnitude of its impact on our modern thinking (Mayr 2004).

As it will be discussed in more detail later on, a noteworthy feat of Darwin's evolutionary theory is in it being able to explain the process of variation in the living world without the need to resort to divine interference in the whole process (Mayr 2001). Notably, the word *evolution* is never used in the seminal book (Ruse 2008). In other words, "Darwin's revolutionary proposal was, thus, to replace the divinely controlled world by a strictly secular world, run according to the natural laws" (490). Likewise, Darwin not only found a way to "describe the patterns that may be found in the history of life on earth, but also to characterize the processes that produced them" (Sober 1984, 13). And natural selection—Darwin's most original and important concept (Mayr 2004)—is one such process that can cause evolution. In more Darwinian terms: "natural selection is *one* process postulated to be capable of producing descent with modification" (O'Grady 1984, 567, our emphasis).

Not surprisingly, evolutionary principles have become a hallmark of modern biology education (Jenkins 2009) so that "a firm grasp of evolution is [considered] invaluable for understanding our own species in addition to the rest of the biological world" (O'Brien, Wilson, and Hawley 2009, 445). At the same time, it is also reported to be one of the most commonly misunderstood and complex areas in the field (Aleixandre 1994; Mead, Hejmadi, and Hurst 2017; Nadelson 2009; Nehm et al. 2009; Penteado, Kavalco, and Pazza, 2012). Even those students and teachers who claim to abide to evolutionary thinking often demonstrate little understanding of its basic principles, including natural selection (Abraham et al. 2009). This, in turn, not only hinders the efforts to teach and learn this foundational theory in biology, but it can also contribute to heighten historical tensions between the science curriculum and individuals' religious belief systems in the classroom (Guessoum 2018).

Drawing on literature in the fields of science education and the analysis of selected classroom-based events, science textbooks and curriculum documents in Canada and Brazil, the present chapter seeks to contribute to the conversation about some of the existing controversy surrounding the teaching and learning of evolution (Friedrichsen, Linke, and Barnett 2016). Here, we suggest, distorted views of evolutionary theory may result from a faulty (and hierarchical) metaphorical understanding of *evolution-as-progress*. That is, the orthogenesis-like (Rudolph, and Stewart 1998) notion that evolution is directed toward known and increasingly complex (superior) biological and social states of existence, which are consequently deemed acceptable (natural and normal) and desirable (adequate). Conversely, those modifications that seem random, unintentional, unpredictable, unplanned, or chaotic are considered to be unnatural, abnormal, and undesirable—not to say disadvantageous, obsolete or inferior—from a socio-biological perspective.

In order to help our readers understand more clearly our line of argumentation, we explore two examples next: one coming from a secondary science classroom and the other from outside the school environment. While we do not claim that they are strictly linked to the idea of evolution as being the ever increasing enhancement of organisms, they nevertheless typify some of the general pitfalls of adopting standards of naturalness, normalcy, and adequacy to assess difference and diversity in life, specially social beings like humans.

It is the summer of 2013. The setting is a private high school in one of the northeast states of Brazil. A senior year student submitted a complaint to the local school board about an image in his physics textbook that he found to be discriminatory (Alves, and Vasconcelos 2013). The illustration (figure 10.1) is meant to demonstrate the concepts of repulsion and attraction between charged atomic particles—that is, protons and electrons.

In order to explain the concept of repulsion, it shows two boys representing positive charges and looking away from each other as if avoiding a possible encounter. Right below, one can see two girls representing negative charges and being *repulsed* by one another. Similarly to the boys, they too look away from one another. In each of these characterizations, arrows pointing in opposite directions reinforce the idea of repulsion or incompatibility. Further down the page, however, one can see a boy and a girl, only this time they are looking at each other as if mutually attracted. In this last case, the arrows that accompany the little cartoonish characters point toward one other.

Although a number of plausible interpretations of the meaning(s) of this image can be drawn, it nonetheless seems to suggest that it is based on a

Figure 10.1 Reproduction of a page of a Brazilian high school physics book depicting boys and girls—and their presumably biological reproductive-oriented affinity—to explain repulsion and attraction of charged atomic particles. The words "repulsão" and "atração" mean repulsion and attraction, respectively. (Source: Maria-Luiza Reis, 2019).
Courtesy of Reis, M.-L. (2019). Representation of an image from a Brazilian high school physics textbook. Ottawa: Private Collection.

strict heteronormative social-biological understanding of sex, sexuality, and reproduction (Snyder and Broadway 2004). Put differently, one of the possible messages it sends to teachers, students, and parents can be the following: much like the natural repulsion, which has been demonstrated by science to occur between identical magnetic poles or particles charged with the same electrical charge, it would be inadequate (undesirable) for individuals (children) of the same sex to attract. That is, it is only natural or normal (acceptable)—within some accepted norms of socio-biological behavior—for boys to feel sexually attracted to girls and vice-versa.

In addition, one cannot ignore the fact that the boys are associated with the *positive* symbol (+) while the girls are represented by the *negative* (-). In terms of their current cultural meanings, those mathematical characters suggest opposite qualities—for example, good/bad, desirable/undesirable, superior/inferior, advantageous/disadvantageous, et cetera.

It is now the summer of 2017. The setting is another Brazilian city, Rio de Janeiro. A couple receives a call from a relative who is looking after their condo while they are vacationing away about a letter that was left on one of their apartment windows (Willmersdorf 2017). Aside the obvious hatred-filled and misguided religious content of the poorly written two-page document (intentionally not displayed here), those who wrote it used scientific-esque argumentation to support their plea for the gay couple to move out of the apartment building where they live. Similarly to the case of the physics textbook image, this letter symbolizes an heteronormative social-biological understanding of sex, sexuality, and reproduction. Had nature created (likely through evolution and natural selection) living beings that could be categorized into an alternative "sexual classification" (e.g., other than female and male), the content of the letter would have been different. Had science revealed a "chromosomal discovery" that justified "the existence of a third sexual behaviour," the presence of the couple in the building would not have been questioned by some of their neighbors.

In the two cases above, it is not possible to accurately determine where the basis of their apparent rejection of homosexuality comes from. It may stem from a combination of misplaced religious orthodoxy, family values, school experiences, et cetera, which makes it impossible to pin down on one place or experience alone. In any case, the attempted entanglement of science and sexual behavior in both instances points to a probable conformity/adherence/acceptance to a set of specific socio-biological benchmarks for what is (or should be) considered natural, normal, and adequate in our society. And, we argue here, these ideals are ingrained in specific occurences within the context of evolution education, specifically through the evolution-as-progess metaphorical thinking. Furthermore, and as contradictory as it may sound, this evolution-as-progress metaphor resembles non-evolutionary (i.e., creationist and intelligent design) claims in that it suggests that evolution and natural selection are purposefully directed/determined toward the perfecting (improvement) of living organisms, especially humans. Likewise, it put humans on the center of everything (especially creation)—thus opposing the idea that we "[are] not outside of the rest of the living world, but [are] actually part of it" (Mayr 2004, 87). In the end, we articulate the implications of our discussion for helping to make school science a site for resistance against prejudice and discrimination.

CONTEXTUALIZING OUR ARGUMENT: DARWIN'S DILEMMA AND THE BIRTH OF CREATIONISM

The social and scientific controversy surrounding the theory of evolution is as old as the idea itself. Even Charles Darwin hesitated for decades before publishing his masterpiece due to the public furor that he knew would follow. Consequently, he struggled with anguish over the obvious backlash that was to come because of the shattering implications, specially to the rigid religious beliefs that were held by a near totality of his British compatriots in the nineteenth century. Among the most notable of implications would be that humans would be removed from their pinnacle of importance among living beings and become reduced to mere animals, having evolved alongside their other animal cousins, including the apes. His theory of evolution by natural selection would shake up the very fabric of his society by challenging the immutable notion that their natural world was made up of God-given organisms for the purpose of human dominion.

Prior to Darwin's work, the general consensus on the nature of the earth and its organisms among religious scholars and most scientists alike was based on the Aristotelian view of purpose and hierarchy in nature (Durant 1998). His (teleological) descriptions of nature as being a combination of form and function nested well into the monotheistic views of the Western world in that the specificity and complexity seen in the world's processes and organisms were divine creations that served a specific function. For example, the rain comes in order to make the grass grow, the long neck of a giraffe is meant to give it access to its food high in the trees, and so forth. In addition, Aristotle's *Scala naturae* (or Scale of Nature), which was created in order to rank all of the earth's organisms into a linear hierarchy, from simple to most complex, fits very well with the religious thoughts on immutable creation and divine purpose among God's living things (Durant 1998). Unsurprisingly, this chain of beings listed plants and fungi at the bottom, followed by invertebrates, then vertebrates upon which humans sat at the apex, just under the angels and God himself. This hierachical and linear view of evolution continues to be present in modern science textbooks whenever images of timelines or trees depicting evolution (especially human) are organized in a way that suggests progressive evolutionary advancements over time (i.e., organisms at the far right side of the picture are equipped with the latest superior evolutionary traints) (Quessada, Clement, Oerke, and Valente 2008).

However, by the late eighteenth century, there was already the brewing of an idea that the world was much older than the accounts that were read from the Bible, as well as the notion that forms of life were dynamic over time, not static, having undergone change through a slow and almost imperceptible process. Geologists like Darwin's close friend the Scotsman Charles Lyell were describing the layers of earth as having been laid down through natural processes, like sedimentation and volcanic activity over millions of years, not the six thousand that was otherwise common religious knowledge at the time (Gillespie 1979). The French anatomist Georges Cuvier had postulated the concept of extinction and of there having been different *kinds* of animals through his studies of fossils, noting the transitional forms that appeared similar to modern-day animals but that did not exist anywhere anymore. It was also a time during which many inconvenient facts were arising through globalization and the colonization of the new world, such as the presence of humans in North America or Polynesia, people whereof the old scriptures did not provide any explanation. Furthermore, the presence of similar species of animals and food plants in the New World to those that were well known in the Old World were not accounted for in the religious documentation and it was becoming increasingly difficult to use religion as a complete and unifying explanation for the observations recorded of natural world.

In light of this context, the response to Darwin's publication in the mid-nineteenth century was mixed: his book quickly sold out of its first printing and it became the topic of vigorous public debate, derision, and ridicule, by scientists, clergymen, and lay-people alike (Mayr 1964). Among scientists, the main points of debate in the early years were centered not so much around the principles of the theory itself, but rather its mechanism (Numbers 1992). Prior to the discovery of the genetic work done by the Austrian monk Gregor Mendel several decades later, there was no understanding for the mechanism of the hereditary nature of the traits subjected to Darwin's natural selection. Much of the rigor of the scientific discourse up until the late nineteenth century was focussed on discussion of the integrity of the blending mechanism of heritable *gemmules*, which was postulated by Darwin in 1859, as well as the comparison and contrast of the natural selection hypothesis to the Lamarkian one of descent with acquired characteristics (Bowler 1988). In 1809,

Jean-Baptiste Lamarck had proposed that the mechanism of change was due to the acquired characteristics in the parent being passed on the offspring, such that a muscle-man like Arnold Schwarzenegger should produce strong, muscular babies that had inherited his athletic physique, for example. A more easily testable hypothesis (and one that came to be used by the German evolutionary biologist August Weismann (1834–1914) to disprove and discount Lamarkian evolution) is that if you cut off a mouse's tail, it should produce pups that also have no tail. It turns out that this is not (always) the case (Sullivan 1981).

Gradually, through further debate and much experimentation, it was shown that Darwinian selection was the fundamental biological process of nature and, in combination with Mendelian genetics, evolution became a core principle of the study of biology by the early twentieth century (Jenkins 2009). The combination of Darwinian and Mendelian biology was then bolstered in 1953 with the discovery of the DNA molecule as housing the building blocks of genetic material and the theory of evolution was concretized as being the fundamental process that underlies life processes. As put by the science epistemologist Theodosius Dobzhansky (1973, 125): "nothing in biology makes sense, except in the light of evolution." That is to say, we can study living organisms and describe their forms but without viewing them through the lens of evolution, we cannot explain how or why they became that way.

The religious response to Darwin's theory of evolution was more complicated. Where religious objections existed, they were usually centered around the dispute between science and the notion of a biblical, God-created, six thousand-year-old earth, that according to the theological doctrine of Special Creationism, housed an array of plants and animals that appeared more recently in their final and current form. Both of these notions conflicted with all of the evidence coming from the astronomical Big Bang to the fossil record of life on earth and all other sciences in between, such as physics and biology. Gradually, the religious leaders were forced to adjust to the growing realization that the earth was much older than was previously thought and that the universe was dynamic, rather than static, therefore change over time was a fundamental aspect of nature (Roberts 1988). Throughout the following decades, the rigid religious doctrines found in much of Europe began to acknowledge that the Bible was not a definitive reference for studying the natural world but was rather more of a guide for the personal relationship between individuals and their God. By the late nineteenth century, various forms of religious churches approached their interpretations of evolution differently, such as Presbyterianism in Scotland embracing evolution wholly, whereas the Irish Presbyterians rejecting it entirely (Livingstone 1999). Alternatively, other forms of religion, such as those adhered to by Catholic and Christian believers, sliced the theory in two, by accepting the concept of evolution, but rejecting the process of natural selection; thus retaining the ultimate cause of the observed patterns to be by the hand of an almighty and intervening God rather than a developed mechanism embedded intrinsically into the fabric of nature.

Outside of Europe, the encounter between Darwin's ideas and religious orthodoxy was no less convoluted. For instance, in the United States, the exisiting decentralized and fragmented educational system allows individual states to determine their own curriculum without much top-down interference from a federal government (Scott 2004). Thus, what is being taught in US schools is often determined by local school boards, which may be populated by individuals who do not necessarily have any expertise in education, causing local values or interests to prevail over more widely accepted notions in many instances.

Furthermore, the founding of religious Fundamentalism in the southern United States in the early 1900s led to a situation in which there was an organized and concerted effort there to eradicate the teaching of evolution in schools and led to the drafting of laws forbidding evolution curriculum from science classes in secondary school instruction. The most famous of these laws was the Butler Act, passed by the state of Tennessee in 1925, which became the subject of the Trial of the Century during which a young high school science teacher John T. Scopes was charged with the accusation of teaching evolution in his science classes.

Even though the prosecution was seen to be weakened by the fact that they had difficulty finding scientist witnesses who would go on the stand to reject evolution as well as the prosecutor's public failure to be able to demonstrate that passages from the Bible could be taken as literal truths, Scopes was found guilty of breaking the law (even though he hadn't yet got to that chapter in the course textbook). Eventually, the decision was reversed by the Tennessee Supreme Court on a technicality and it was thrown out. Following this trial, more anti-evolution bills continued to be proposed in other southern states over the following decades, some passed but even more were defeated before becoming law.

As science started playing a much more stable and respected role in 1960s' American society than it had during the Scopes trial in the 1920s, the Fundamentalist religious right sought to add religion to the teaching of life sciences in schools as opposed to attempting to abolish the scientific teachings of evolution outright. However, because the teaching of religious views in high schools was an unconstitutional violation of the separation of church and state, the Fundamentalists devised a way to insert the Biblical teachings into science classes by presenting Creation Science as an alternate, (yet supposedly scientific) view and argued for its inclusion into the curriculum, in order to tell a *balanced* view of *both sides* of the evolution story (Armstrong 2004).

Invented by a hydraulic engineer, Henry M. Morris, Creation Science was presented as a way to explain the Biblical stories via modern scientific methods, such as using geology to explain Noah's Flood and calling it *flood geology*. Creation Science not only adhered strictly to the literal interpretation of Special Creation, with a young universe and earth as well as immutable organisms that were created by God in their current forms, but it also presented itself as the only *rational* alternative to evolution. This latter aspect was to accomplish two goals, the first being to justify its teaching in science classes alongside evolution, but also to create the false notion that inconsistencies or uncertainties in the theory of evolution were to be heralded as support for creationism. Over the following decades, Morris and his Creationist colleagues would form societies, publishing companies, and Young Earth ministries that would evangelically promote his Creation Science views widely throughout US society, eventually garnering much political and financial support from sympathetic politicians and patrons alike (Armstrong 2004).

By the late 1970s, many campaigns were underway in US states to rally the public behind legislation that would allow Creationism to be taught alongside evolution in science classrooms, as a way to maintain a *balanced curriculum*. They argued that it was academically insufficient to teach evolution *without alternative theory of origins*. As a result, there were proposals of legislation put forward in twenty-seven different states by the early 1980s to have creationism taught wherever evolution was also. In all but two states, the *evolution and creation equal time* bills were quashed at the committee level but they managed to pass into law in Arkansas and Louisiana. Both states' laws were quickly challenged by coalitions of science educators, the ACLU, as well as a variety of clergy representatives that argued that the *both sides* law violated the Establishment Clause of the First Amendment (meant to ensure that Americans can be guaranteed their education to be free from the influence of an unwelcome religion) by promoting a religious point of view, that creationism was not scientific and there was no secular reason to teach it (Herlihy 1982). The laws were brought down as unconstitutional and as such, it was no longer legally viable to attempt to teach creationism in science classrooms in the United States, unless it could be rebranded as something that could survive challenges based on the Establishment Clause.

Following the 1987 Supreme Court decision against the Louisiana *both sides* law, the last and the most recent phase in the history of the opposition to evolution in the United States emerged. The idea was first proposed in a book published by authors from a conservative Christian organization in Texas called the Foundation for Thought and Ethics (FTE) that put forward arguments against evolution without mentioning any references to the Bible, such as *creator*, *creation*, or the floods (Numbers 1992). They did not put forward any actual or tangible mechanism, instead basing their argument on the fact that the complexity and beauty in the design of nature is so spectacular that it was unexplainable by natural causes and had to be driven by an *intelligent* force. Thus, Intelligent Design (ID) became the final throe in the movement to insert creationist religion into science classrooms.

Intelligent Design is based on the concept of *Irreducible Complexity*, which implies that only an "intelligent" force could create the smallest living units that still require multiple parts to survive. As a metaphor for the complexity of a living cell, the ID proponents argue that it is like a mousetrap that wouldn't work anymore if you remove one of its parts, therefore biological evolution by natural selection couldn't possibly explain the origin of a living cell, by their claims. However, the modern scientific understanding of biological evolution is not limited to describing the cause of changes in higher organisms over time, but it has also managed to explain the evolution of molecules, organelles, and cells, not to mention the evolution of organs, organ systems, and even behavior. As a result, the incredulity argument against evolution in Irreducible Complexity is easily disproven and therefore Intelligent Design cannot be considered a rational scientific alternative to evolution by natural selection.

Furthermore, it doesn't even satisfy the definition of science itself, which would be a study of the natural world by proposing falsifiable claims that can be tested using natural, material means. The whole premise of ID is that it must rely on the power of an *intelligent* force from a supernatural origin, one which cannot be observed, measured, verified, or falsified. Invariably, exposure to this *alternative* explanation has had its impact on the understanding of science among Americans, roughly half of whom reject the theory of biological evolution entirely to this day. Yet, due largely to the social and political attention given to it, Intelligent Design manages to survive in much of the US curriculum on evolution, propping itself up as a supposed scientific alternative to evolution based on a combination of untestable claims and by ignoring scientific evidence that negates its claims that can be tested.

Over the past decades, the growth of religious fundamentalism has also spread from the United States to other countries, including Brazil (Dorvillé and Teixeira 2015) and Canada (Barker 2004), which has also sparked the *equal time* debate over whether creationism should also be taught in science classes in these countries. Although, the views toward education in these countries may differ somewhat, in that acceptance of Intelligent Design hypothesis is high in Brazil (Silva et al. 2016) and not so for Canada (Ashgar et al. 2015).

A recent Angus Reid poll (2019) showed that 61 percent of Canadians think that earth's organisms evolved over long periods of time and only 22 percent correspond to the creationist ideas. Despite this strong support for evolution, Canada has experienced its own history of creationists attempting to bring Intelligent Design into the science classrooms. Most notably was the dispute between a regional school board in Abbotsford, British Columbia, and that province's Ministry of Education in 1995 over that Bible Belt community's attempt to uphold a policy of teaching biblical interpretations of the earth and its organisms in the science classes of its schools (Barker 2004). That isolated conflict resulted in the Abbotsford school board being told that religious views could not be taught as science and that it had to uphold the provincial curriculum, including the teaching of evolution and that teachers could not opt out of doing so.

Very few science educators—if any—would disagree with the perspective that the science classroom is not the place to teach religion. However, creationists believe that the teaching of evolution is not a neutral position, but rather that it promotes anti-religious ideology. In response to this perceived threat, creationists started using language, such as calling the evolutionary biologists *evolutionists* that *believe in evolutionism*, in an attempt to bring evolution down to the level of being just another biased belief system, on par with their own. Another common problematic use of wordplay involves the creationist argument against evolution being *just a theory*, which seems to infuse a level of uncertainty and speculation to the scientific notion (Coyne 2009). This effect rests on the distinction between the scientific and colloquial definitions of the word, wherein the lay usage may imply that aforementioned speculative meaning, but the technical use of a *scientific theory* is that it refers to mountains of inter-related and mutually supported evidence that has not been disproven. As such, it is essentially a scientific fact. By comparison, the ideas behind Theory of Special Creationism or of Intelligent Design, which are mostly untestable and non-falsifiable statements would otherwise go by the distinction of being a series of hypotheses. Insomuch as some of the tenets of Special Creationism are testable (age of the earth, Noah's flood, immutable forms, etc.) or for the claims made in support for Intelligent Design, these hypotheses have been tested against empirical evidence and have been disproven, thus falsifying those claims and refuting their validity as fact.

EVOLUTION EDUCATION IN BRAZIL: PUBLIC POLICIES, CURRICULUM DOCUMENTS, AND SCIENCE TEXTBOOKS

Brazilian educational legislation is regulated by the Lei de Diretrizes e Bases da Educação (Law of the Guidelines and Foundations of Education), known as LDB, which provides for the establishment of a curricular basis for the contents to be addressed in basic education. The first attempt to establish this curricular base was the Parâmetros Curriculares Nacionais (National Curricular Parameters or PCN in Portuguese [Brasil 1998]) and later, at the request of high school teachers, a new (complementary) version was elaborated with greater detail of the contents to be addressed in the classroom: PCN+ (Brasil 2000). These documents establish parameters and guidelines for approaching all subject areas of elementary education. More recently, a new document was developed to guide the teaching and learning of school content subjects. This document is called the Base Nacional Comum Curricular (National Common Curricular Base or BNCC in Portuguese). It was initially prepared for pre-school and elementary education and later for secondary education (Brasil 2018). Thus, as the country is still transitioning from the PCN/PCN+ to the BNCC, we looked at these three official Brazilian documents for our analysis of the recommended approach to evolution in high schools. On that note, we found the PCN (1998) to contain the most significative quotes for our argument.

In searching the three documents for instances where the term *evolution* is mentioned, directly or indirectly, one will find three convergent aspects regarding the approach to evolutionary contents: (1) evolution as a structuring axis of biological knowledge; (2) the interdisciplinary approach in the study of evolution; and (3) the transience of scientific knowledge. The excerpts of the documents containing these aspects will be transcribed and discussed below.

The famous Dobzhansky's phrase is represented in all the documents analyzed. In several situations, the knowledge about the evolutionary processes are indicated as fundamental for the understanding of the other contents of biology. For isntance: "For the study of the diversity of living beings, traditionally of Zoology and Botany, the evolutionary-ecological approach, that is, the geological history of life, is adequate" (Brasil 1998, 18).

Through similar passages that exist in the documents analyzed, we can see that they recognize and highlight evolutionary processes as fundamental for the teaching of various contents of biology, especially those related to living beings and the environment. Consider the evolutionary aspects when addressing taxonomy, anatomy, and comparative physiology, as well as ecology can avoid the vision of evolution as progress or teleology (Reiss 2009).

At some points in the documents the importance of other areas (e.g., chemistry, physics, and geology) is emphasized for the understanding of the evolutionary process. For example:

> Disciplinary learning in biology, whose setting, the biosphere, is an articulated whole, is inseparable from other sciences. The very understanding of the emergence and evolution of life in its various forms of manifestation demands an understanding of the geological and environmental conditions prevailing on the primitive planet. (Brasil 1998, 9)

It comes with no surprise that the interdisciplinary character of the study of evolution is present (acknowledged) in the documents: in the construction of evolutionary theory, Darwin used elements of geology that were based on the findings of Charels Lyell (1797–1875). Later discoveries in several other areas helped to corroborate Darwin's ideas. The signaling of this multidisciplinary integration in guiding documents for education indicates the importance of collaboration and cooperation for the construction of scientific knowledge.

In the three documents, the transitory (nonabsolute) nature of scientific knowledge is emphasized, and specifically about evolution, the divergence of ideas, the breaking of paradigms and historical aspects are highlighted. For instance:

> To know some explanations about the diversity of species, their presuppositions, their limits, the context in which they were formulated and in which they were replaced or complemented and reformulated, allows the understanding of the historical-philosophical dimension of scientific production and the character of scientific truth. (Brasil 1998, 17)

It is important to show students that scientific knowledge is built on the discussion of ideas that may be convergent, complementary or even divergent. These are the result of research in certain historical and social contexts, which at a given moment can be accepted as *good* (sufficient) explanations for certain phenomena and which can later be refuted. Such is the case with evolution.

In Brazil, school textbooks are delivered to public schools through the Programa Nacional do Livro e Material Didático (National Book and Instructional Material Program or PNLD in Portuguese). The books can be chosen by the teachers, from a list previously organized by specialists and made available digitally: the Textbook Guide. This being a federal gorvernment program that distributes books free of charge, this initiative is essential for Brazilian public schools. In the case of biology textbooks, they have been distributed to public high schools since 2007.

Among the objectives of the PNLD, we would like to highlight the following three: (i) to improve the teaching and learning processes in elementary public schools, with the consequent improvement of the quality of education in general; (ii) to guarantee the quality of the materials to support the educational practice in elementary public schools; and (iii) to support the autonomy and professional development of teachers (Brasil 2018).

The teaching of evolution in Brazilian textbooks has already been an object of study in previous works, which in general indicate problems in the approach, conceptual errors or possibilities of misinterpretation of evolution contents. Some of this research will be described below.

Almeida and Da Rocha Falcão (2010) point out that the theme of evolution has been approached without historical contextualization in textbooks, which may make it difficult also for the stduent to understand how related concepts have developed over time. In that sense, the use of the history of science can help in the understanding of ideas, showing the student that "the gradual and slow process of knowledge construction [allows for] a more concrete view of the real nature of science, its methods, its limitations" (Martins 1998, 18). In working with the history of evolutionary theory, the teacher can deconstruct with his student the image of a neutral science, impartial and without link with society, presenting to this science as a human activity that has no answer to all things and that is built collectively (Azevedo and Motokane 2011).

Moody (1996) states that the content of evolution should not be treated as a mere theme because it is a theoretical foundation of biology. However, the way in which the content is presented in textbooks does not treat evolution as the central axis of biology since they do not use it as an organizing thread (Tomotoni and Salvador 2017). For Azevedo and Motokane (2011), it would be important for the theme of evolution to be a unifying principle in the curriculum so that it would combine different areas of biology, thus explaining the origin and diversity of life as well as its functioning.

According to Bizzo (1991), textbooks present students with a simplistic version of the development of biological theories as well as versions no longer accepted in the academic environment. As a result, they face obstacles to understanding organic evolution due to the simplification (knowledge transposition) that evolutionary theories undergo (Almeida and Da Silva Falcão 2010). This simplification may end up constructing misconceptions not only for the students, but also for the teachers who make use of this material in the planning of their classes. For example, the textbooks may present Lamarck and Darwin as *mentors* of the term, when in fact they have never used it (Belinni 2006). Besides, it claims that Lamarck used the term *transformism*, believing that organisms were capable of change, while Darwin used the term *descent by modification*. The term evolution would have been coined by the

German biologist Albrecht Von Haller, who conceived the theory that the embryos would have developed from preformed homunculi in the ovum or the sperm. Accordingly, as Von Haller believed that the embryos *evolved*, the term evolution has been since commonly associated with a sense of *progress,* a notion always fought by Darwin (Belinni 2006).

In general, textbooks feature Darwin as a model scientist while Lamarck is no more than a speculative theoretician (Almeida and Da Silva Falcão 2010). In addition, Lamarck's theory as presented in the textbooks is quite distant from the one formulated by him. Belinni (2006) points out in the textbooks, Lamarck is linked to the concept of inheritance of acquired characters, which was not originally his idea. Indeed, the processes of transmission of characteristics of living things are known to exist since the eighteenth century. For Dias and Bortolozzi (2009), textbook authors often times approach the Darwinian theory without any criticism, not pointing out its scientific weaknesses as currently discussed by experts in the field.

What Science Textbooks and Official Curriculum Documents Have to Say

Studies that analyze the approach of high school science textbooks to the teaching and learning of evolution are extremely important because they contibute to the avoidance of the perpetutation of conceptual errors. Otherwise, students would be reaching erroneous conclusions on topics related to evolution, such as mutation, variability, extinction of species, among others. However, since there is a program in Brazil that provides free books for all public schools, with a significant expenditure of public financial resources, we ask the following question: Do these books contemplate the curricular guidelines present in the official documents? Based on this question we analyzed two collections of wide distribution in Brazil: *Modern Biology* (Amabis and Martho 2016a, 2016b, 2016c) and *Bio* (Lopes and Rosso 2015a, 2015b, 2016c). The analysis was based on the three aspects highlighted initially in the official documents: Evolution as a structuring axis of biological knowledge, the interdisciplinary approach in the study of evolution and the transitoriness of scientific knowledge. Each of these collections has three volumes, which correspond to the three years of high school in Brazil.

Evolution as the Structuring Axis of Biological Knowledge

In the two collections we analyzed, we found a similar approach, where the theme evolution is presented as being extremely important for the understanding of biological knowledge. There were even specific chapters where the topic was highlighted in the books for the first-year high school students. However, the concept of evolution is not taken up in the following chapters. For example, in the chapters related to ecology there is no discussion, from an evolutionary point of view, of the adaptations of organisms or ecological interactions such as predation and herbivory. In subsequent books, the same situation is observed. In the classification of living beings, phylogenetic is a fundamental aspect for the current classification system. However, in the textbooks, organisms are presented in a linear way without dealing with ancestry, divergences, convergences, or adaptations. In the last volume of the two collections there are specific chapters for evolution, where important concepts are presented, but it is not possible to identify that the evolutionary theme permeates all fields of biology (covered in the previous volumes).

The Interdisciplinary Approach in the Study of Evolution

In the *Bio* collection the authors in many situations demonstrate the importance of knowledge of other areas for the understanding of evolutionary processes. When the authors present the origin and evolution of the first organisms, they bring geology in to subsidize the discussion, by relating some adaptations of organisms with climatic or geographical aspects, but they do not call attention to the processes involved. It is as if there was only one aspect worth learning about: the *final product.*

In the collection *Modern Biology*, the authors sometimes suggest possibilities for an interdisciplinary approach, but nothing specific to the teaching of evolution as foreseen in the legislation. They simply present a chapter on human evolution, where they relate evolution and culture and show the impact of culture on the evolution of humanity, especially from the industrial revolution.

The Transitoriness of Scientific Knowledge

Although in both collections the authors present the various theories related to evolution throughout history, demonstrating the transitoriness of knowledge and the contribution of several studies to the current understanding about evolution, this presentation is linear and devoided of any social and historical contexts. The tensions and conflicts that have driven many of these contributions to science are not presented.

In addition to the problems already identified by other authors about evolution teaching, we can see that the authors of high school textbooks do not fully follow the guidelines presented in the official documents for the approach to evolution in Brazilian school. This situation at first may seem like just a choice of different points of view, but it can have a significant impact on classroom actions. If we consider that official documents guide the construction of state and municipal curricula, which are used by teachers to plan their classes, and teachers receive books from the federal government that do not follow these guidelines, we will have a contradictory situation for teachers to solve. Certainly, the textbook should not be the only source for the planning of teachers, but in a country of dimensions such as Brazil, with extremely variable socioeconomic realities, with programs of teacher training of variable quality, we will find in many situations the teacher choosing the easier and gives you more security, the book will be the basis of the curriculum in the classroom. Issues such as this should be considered in the design and implementation of public policies such as PNLD and BNCC, as they should be in line with the expected objectives.

EVOLUTION EDUCATION IN CANADA: IS IT REALLY THAT DIFFERENT IN THE GREAT WHITE NORTH?

In what follows, we briefly report on four incidents that are distant in time and space, but close to one another in their significance to our argument. We invite our readers to try to deduce what the connection among them might be and how they fit into our argument and the theme of the book. In addition, the recounting of these experiences is offered here as an opportunity for people to (re)consider how evolution and natural selection can be perceived outside of the Brazilian cultural milieu. By no means do we intend to propose that our examples are representative of the entire Canadian educational reality (the same way we do not claim to have offered a comprehensive examination of the public policies, curriculum documents, and science textbooks in Brazil). Yet, the two events described below offer grounds to consider that our argument is not bound to geographic borders.

It is the winter of 2010. The setting is a science methods class for high school biology student teachers at a Canadian university. One of the course requirements is for groups of students to demonstrate how to effectively teach a lab activity. As part of their usual presentation protocol, students provide a short theoretical groundwork before proceeding to the hands-on aspect of their demonstration. The presentation on cell division seems to be going well when one of the members remarks that Down Syndrome as an example of when meiosis *goes wrong*. That last sentence took the audience aback. Would it have bothered you?

It is the early 2010s. Peter is a zoo interpreter who is frequently invited to do presentations to science students in the teacher education program at a

Canadian university. He works for a local, privately funded zoo and animal rescue organization. The talks are usually one hour long and include interactions with live animals—for example, crocodiles, snakes, turtles, and so forth—that are properly restrained when necessary; for example, the crocodile's mouth is duct taped shut to avoid biting. Among the various performances that Peter delivered, we have selected one quote that is relevant for our argument:

> Now, crocodiles are probably the best creatures to use as an example when we're talking about evolutionary characteristics because everything about them has evolved for a reason, everything from where their nostrils are to their eyes on their body to the plates of bones in their skin, their hands and feet. Everything helps them, contributes to their survival.

Based on Peter's discourse, what would you say is his understanding of the concept of animal evolution? Do you agree with his view?

What Is Going on When a *Wrong Person* Is Born and Reptiles Evolve?

Could it be that what the student said (even so inadvertently) in the first event shocked people in the room because it suggested that individuals with Down Syndrome (forty-seven chromosomes) are the result of an error? Could it be that he also gave the impression of calling those individuals a mistake? Regardless, what transpired that day as the result of the student's remark on the process of meiosis is not unprecedented (Reis, Ng-A-Fook, and Glithero 2015). Biology textbooks commonly found in Canadian university libraries are filled with the same depreciating and prejudiced semantics used to describe Down Syndrome (DS). For example: "errors and exceptions in chromosomal inheritance" (Campbell et al. 1999, 271), "non- disjunction disorder" (Miller and Levine 1991, 235) and "abnormal meiosis" (Ritter et al. 1993, 556). The words *syndrome* or *mutation* themselves are synonyms with *anomaly*. Although this language is (probably) meant to indicate that the cell division did not generate a faithful progeny, one must question if it does not equally insinuates that individuals carrying that type of trisomy are an unacceptable (unnatural and abnormal) and inadequate (undesirable) biological product, one that probably should have been eliminated by natural selection during the short evolutionary history of our species. Why else would DS be considered such an oddity?

At the same time, and as an example of a subject-specific school document, the Ontario high school biology curriculum mandates that its implementation "must, wherever possible, be inclusive and reflect the diversity of the student population and the population of the province" (Ontario Ministry of Education 2008, 16).

Put differently, teachers can choose how to best meet the provincial curriculum goals as long as they do not lose sight of their students' existing socio-cultural and cognitive (biological) multiplicity. Strangely enough, the curriculum document has twelve instances where the word *meiosis* appears, two where the word *trisomy* is used, but none for Down Syndrome. Despite this, DS remains part of the fifth most common "developmental disabilities or disorders" between children aged 5–14 in Canada (StatCan 2001), which is one of the signing countries to the UN's Convention on the Rights of Persons with Disabilities (2006).

As for Peter, he offers explanations for the existence of specific characteristics (behavioral or anatomical) shown by certain animals in the wild. His discourse alludes to the notion that along animals' evolutionary trajectories, natural selection has afforded them what they needed/wanted to survive. That being so, "everything about [crocodiles] has evolved for a reason. . . . Everything helps them, contributes to their survival." Other examples of this line of reasoning (not shown here) include to say that reptiles that have developed the ability to give birth to live young are "generally considered to be a little more evolved" because their babies are able to survive better and also the perception that certain species of snakes developed the ability to eat other snakes presumably out of their "desire to eat" their competition.

In all those instances, Peter's language arguably evokes a sense of directionally to the notion of evolution and natural selection. In that sense, the animals' *desire* to live seems to be sufficient *reason* for changes to occur—be them reproductive, anatomical, behavioral, and so forth. Conversely, any changes that are without purpose or intention can be dismissed unnatural, abnormal (unacceptable) and inadequate (undesirable). Here, we question if what Peter, as a public educator, said could influence how biology is going to be taught by his audience of future teachers once they graduate.

Both cases are similar in their core as they subscribe to the metaphor of evolution-as-progress. That is, each situation is characterized by a general acceptance of presumed standards of naturalness, normalcy and adequacy. In other words, what is natural (or intended by nature) is also normal (acceptable) and adequate (desirable)—and vice versa. So, there is an appropriate way for cell division to occur and the acquisition over time by certain reptiles of the proper anatomical and behavioral characteristics to secure their existence. (Similarly, as exemplified in the introduction, there is also the assumption that men and women must conform to a correct way to engage in relationships with one another). Only when these (natural and desirable) expectations are met are these individuals accepted as *fitting* (normal and adequate) (Mills and Beatty 2006).

We believe that the artifical standards of naturalness, normalcy, and adequacy—as presented here—were incorrectly established through a faulty understanding of the concepts of evolution and natural selection. In such cases, those ideas would be seen as constantly working not only toward the perpetuation (maintenance and growth) of species, but also their constant progress (amelioration). This is the case with people who are more or less accepted by other members in their community or animals that can be classified as more or less *evolved*. In the end, notions of (un)naturalness, (ab)normality, (in)adequacy are all imbued in this notion of natural selection as a biological mechanism for progress, meant to keep the evolution of species on track to an ultimate ideal(ized) existence.

The idea that "any system undergoing an evolutionary change must be subject to a steady increase in its state of complexity—[that] evolution *must* in some sense be a 'progressive' change" (Bowler 1975) is problematic to say the least. Worst: it abides to similar values put forth by theories like creationism and intelligent design, which acknowledge intentionality—or directionality—in the original formation and continuous expansion of the universe as we know it. That is, in those instances where the concept of evolution is generally associated with the betterment of species to guarantee the survival of the fittest, little distinction can be made between evolutionism and its opposing theories. This, in turn, creates the illusion that certain biological (physical and cognitive) traits and social behaviors are not only universally desirable over others, but also *normal* or *natural*. More dangerously: it can contribute to a sense that science (specially as it is taught in schools) is a linear process free of controversies. Therefore, to make our schools a space of resistance to counter the notion that Darwin's theory explains biological and social progression as well as embraces ideals of directionally to perfection is a way to promote a teaching of evolutionism that is both scientifically sound and anti-discriminatory. Ultimately, we contribute to the overarching theme of the book by arguing that evolution education can take place in multiple spaces (in the media or science classrooms) and by many actors (teachers, public educators, and anonymous individuals). More importantly, we must be mindful of the real possibility that evolution can be—as it is often the case—taught and learned in similar ways as creationism and intelligent design. (The expression *wolf in sheep's clothing* comes to mind).

FINAL THOUGHTS: WHY METAPHORS MATTER?

Discourse is a crucial means of communication used by people during the course of their activities, of which school is but one of them (Reis 2007). Aside from other actions like gestures and the use of tools available in the environment (e.g., books), much of what is accomplished by people in schools is made available to interested observers through the language used—or not—in their discourses. According to Bowers (2002), the language we choose to use reproduces the dominant cultural group's moral norms that govern our atributes and relationships with one another and the world.

That is to say that the language employed by people and found in science textbooks and curriculum documents carries a message on its own and represents the historical footprint of those social, scientific, political, cultural, and so on, battles of interest in society (Reis, Tippett, and Scott 2014; Knain 2001). Therefore, language use is the result of the struggle for the predominance of a single ideology over others and one should not think of it as being accidental. Language use has no face value.

Since the mid-twentieth century philosophers have accepted that metaphors permeate all discourse and are fundamental to human thought (Aubusson, Harrison, and Ritchie 2006). Metaphor is not merely a linguistic phenomenon, but also a fundamental principle of thought and action (Lakoff and Johnson 1980), one that structures our systems of knowledge and belief in a pervasive and fundamental way (Fairclough 2006), including our understanding of evolution and the challenges around its teaching and learning.

Metaphors are the product of a linguistic choice, so that "when we signify things through one metaphor rather than another, we are constructing our reality one way rather than another" (Fairclough 2006, 195). Consequently, we become "responsible for the metaphors we choose to privilege and we need to be critically (and self-critically) responsive to the effects of their deployment" (Doll and Gough 2002, 74). That said, we invite science educators to address the ways in which language can foster the detrimental perception of evolution as progress. Otherwise, we run the risk to ignore an existing double blind (Bowers 2001) in the field: the fact (contradiction) that certain scientific ideas that are repeatedly promoted through textbooks and curriculum—such as evolution and natural selection—are populated with metaphors that hinder how the nature of science is publicly perceived.

In the specific case of the object of our argument, the combined interest in the study of metahphors in/for evolution education becomes crucial. After all, etymologically, the word evolution refers broadly to the idea of directional (forward) motion and predestined activity (Hodgson and Knudsen 2010).

Notwithstanding, common sense has that the word *progress*—much like its derivatives, such as *progression*—applies to those new things having come or brought into being, something now existing or appearing for the first time. In our current cultural thinking system, embedded in the technological revolution of our time, what goes without saying is that the new—whatever it might be—is considered to be the better version of a kind. Truth be told, why shouldn't we expect the latest cell phones or cars to be any less sophisticated than the previous models? However, some might argue that our tendency to associate what is new with the idea of improvement is the reason why the word progress is constantly used as a metaphor to explain, justify, and promote change—which, in turn, can lead to an uncritical acceptance of the new as superior. Put simply, the metaphor of evolution-as-progress implies that every modification is positive and necessary, or else it would not have occurred in the first place. What's more: this metaphor denotes a desire for these changes to occur. As a result, and without being a creationist, one is led to believe in the tenacious notion that nature is worth imitating because natural selection generates a continual improvement of species (progress) (Giraldeau 2016).

As we have insistently argued here, the metaphor of evolution-as-progress perpetuates a widespread misunderstanding that percolates realms of our lives beyond schools. If not properly identified and corrected, such misconstriction can favor dangerous ideals of perfectionsim and supremacy in society, of which *social Darwinism* is a classic example. (For a moment, consider the potential problems associated with the fact that social Darwinism was mentioned in passing in only one of all the Brazilian documents analyzed (see page 559 of the BNCC [Brasil 2018]).

Our analyses of classroom-based events, selected current science textbooks and curriculum documents in Canada and Brazil suggest that the sociocultural implications of science in controversial matters like prejudice and discrimination, social inclusion in school and the workforce, economic impact on health care system, effect on family structure, and so forth, are expected to consume little time of classroom instruction. Alternatively, science education has the potential to bring to the forefront the reality that certain categories of naturalness, normalcy and adequacy are not fixed concepts, but ever evolving ideas.

In this chapter, we attempted to rekindle the discussion around the possible ways that science educators can approach the teaching of evolution (as a controversial topic that it can be) by denying any equivalence with progress or biological determinism (Graves 2015) as well as adopting an *ethics of difference* (Koch 2005), where diversity among living beings (specially humans) is inherent to life itself. In doing so, we invite science educators alike to appreciate how language contributes to create and sustain a taken-for-granted description of reality that favours specific attitudes toward particular cultural norms of acceptance (Cox 2010; Wilson 2012).

Although others have come to similar conclusions (Andrzejewski, Baltodano and Symcox 2009; Martusewicz, Edmundson, and Lupinacci 201 Keller 1992), it was Bowers (2001, 414) who we believe probably said it best: "Language is not a conduit for communicating objective knowledge. Rather, it carries culturally specific ways of thinking—and teachers and students are connected, often in unconscious ways, to this symbolic ecology." To deny that reality is to allow the politics of science education (Reis 2010) to continue to shape and validate certain privileged narratives in school science—especially evolution education—that could insist in ignoring the voices of those already marginalized not only by their physical and mental abilities, but also their gender, skin color, socioeconomic condition, age, ethnicity, and immigration status.

REFERENCES

Abraham, Joel K., Eli Meir, Judy Perry, Jon C. Herron, Susan Maruca, and Derek Stal. "Addressing Undergraduate Student Misconceptions About Natural Selection With An Interactive Simulated Laboratory." *Evolution: Education and Outreach* 2, no. 3 (2009): 393-404. doi:10.1007/s12052-009-0142-3.

Aleixandre, Maria Pilar Jiménez. "Teaching Evolution And Natural Selection: A Look At Textbooks And Teachers." *Journal of Research in Science Teaching* 31, no. 5 (2007): 519–535. doi:10.1002/tea.3660310507.

Almeida, Argus Vasconcelos De, and Jorge Tarcísio Da Rocha Falcão. "As Teorias De Lamarck E Darwin Nos Livros Didáticos De Biologia No Brasil." *Ciência & Educação (Bauru)*16 no. 3 (2010): 649–665. doi:10.1590/s1516-73132010000300010.

Amabis, Jose Mariano, and Gilberto Rodrigues Martho. *Biologia Moderna*, vol. 1. São Paulo: Moderna, 2006a.

Amabis, Jose Mariano, and Gilberto Rodrigues Martho. *Biologia Moderna*, vol. 2. São Paulo: Moderna, 2006b.

Amabis, Jose Mariano, and Gilberto Rodrigues Martho. *Biologia Moderna*, vol. 3. São Paulo: Moderna, 2006c.

Andrzejewski, Julie, Marta Baltodano, and Linda Symcox. *Social Justice, Peace, And Environmental Education*. New York: Routledge, 2009.

Angus Reid. 2012. "Britons and Canadians More Likely To Endorse Evolution Than Americans." Accessed April 1, 2019. http://angusreid.org/britons-and-canadians-more-likely-to-endorse-evolution-than-americans/.

Armstrong, Karen. *The Battle For God*. New York: Harper Perennial, 2004.

Ashgar, Anila, Sarah Bean, Wendi O'Neill, and Brian Alters. "Biological Evolution in Canadian Science Curricula." *Reports of the National Centre for Science Education* 35 (2015): 1.1–1.21.

Aubusson, Peter J, Allan G Harrison, and Stephen M Ritchie. *Metaphor And Analogy In Science Education*. Dordrecht: Springer, 2006.

Azevedo, Renato, and Marcelo Motokane. "Evolution In Pre-high School Textbooks Approved By MEC: A Reflection From The Analysis Of Two Cllections." *VIII Encontro Nacional de Pesquisa em Educação em Ciências* (ENPEC), Campinas (Sao Paulo), 2011. http://abrapecnet.org.br/atas_enpec/viiienpec/resumos/R0641-1.pdf.

Barker, John. "Creationism In Canada." In *Cultures Of Creationism: Anti-Evolutionism In English-Speaking Countries*, edited by Simon Coleman and Leslie Carlin, 85–108. Aldershot: Ashgate, 2004.

Bellini, Luzia. "Avaliação Do Conceito De Evolução Nos Livros Didaticos." *Estudos em Avaliação Educacional* 17, no. 33 (2006): 7–28.

Bizzo, Nelio. "Ensino De Evolucao E Historia Do Darwinismo." PhD diss., Universidade de São Paulo, 1991.

Bowers, Chat. "Toward A Cultural And Ecological Understanding of Curriculum." In *Curriculum Visions*, edited by William Doll and Noel Gough, 75–85. New York: Peter Lang, 2002.

Bowers, Chet. "How Language Limits Our Understanding of Environmental Education." *Environmental Education Research* 7, no. 2 (2001): 141–151. doi:10.1080/13504620120043144.

Bowler, Peter J. "The Changing Meaning Of Evolution." *Journal of the History of Ideas* 36, no. 1 (1975): 95–114.

Bowler, Peter J. *The Non-Darwinian Revolution*. Baltimore: The Johns Hopkins University Press, 1992.

Brasil. *Parâmetros Curriculares Nacionais: Ensino Médio*. Brasília: Ministério da Educação—Secretaria de Educação Média e Tecnológica, 1998.

Brasil. *Base Nacional Comum Curricular (BNCC)*. Brasília: Ministério da Educação—Secretaria de Educação Média e Tecnológica, 2018.

Brasil. *PCN+ Ensino Médio: Orientações educacionais complementares aos Parâmetros Curriculares Nacionais*. Brasília: Ministério da Educação—Secretaria de Educação Média e Tecnológica, 2000.

Campbell, Neil, Jane Reece, and Lawrence Mitchel. *Biology*. San Francisco: Benjamin-Cummings, 1999.

Conniff, Richard, and Geoffrey Giller. "Evolution Of An Icon." *Cosmos* 61 (2015). Accessed April 1, 2019. https://cosmosmagazine.com/society/evolution-icon.

Cox, Robert. *Environmental Communication And The Public Sphere*. London: Sage, 2010.

Coyne, Jerry. *Why Evolution Is True*. New York: Oxford University Press, 2009.

Dias, Fernanda, and Jehud Bortolozzi. "How Biological Evolution Is Approached In High School Books." *VII Encontro Nacional De Pesquisa Em Educação Em Ciências* (ENPEC), Florianópilis (Santa Catarina), 2009. http://posgrad.fae.ufmg.br/ posgrad/viienpec/pdfs/670.pdf

Dobzhansky, Theodosius. "Nothing In Biology Makes Sense Except In The Light Of Evolution." *The American Biology Teacher* 35, no. 3 (1973): 125–129. doi:10.2307/4444260.

Doll, William E, and Noel Gough. *Curriculum Visions*. New York: Peter Lang, 2002.

Dorvillé, Luis, and Pedro Teixeira. "The Growth Of Creationism In Brazil: Major Influences And Recent Developments." *X Encontro Nacional De Pesquisa Em Educacao* (ENPEC), Agua de Lindoia (Sao Paulo), 2015. http://www.abrapecnet.org.br/enpec/x-enpec/anais2015/resumos/R0234-1.PDF

Durant, John. "A Critical-Historical Perspective On The Argument About Evolution And Creation." In *An Evolving Dialogue: Scientific, Historical, Philosophical And*

Theological Perspectives On Evolution, edited by James Miller, 265–280. Washington: American Association for the Advancement of Science, 1998.
Fairclough, Noel. *Language And Globalization.* London: Routledge, 2006.
Friedrichsen, Patricia, Nicholas Linke, and Ellen Barnett. Biology Teachers' Professional Development Needs For Teaching Evolution. *Science Educator* 25, no. 1 (2016): 51–61.
Gillespie, Neal. *Charles Darwin And The Problem of Creation.* Chicago: University of Chicago Press, 1979.
Giraldeau, Luc-Alain. *Dans L'œil Du Pigeon.* Montreal: Boréal, 2016.
Graves, Joseph L. "Great Is Their Sin." *The ANNALS Of The American Academy Of Political And Social Science* 661, no. 1 (2015): 24–50. doi:10.1177/0002716215586558.
Guessoum, Nidhal. "Religion In Science Classes: Heresy Or Constructive Pedagogy?" In *13 Questions: Reframing Education's Conversation—Science*, edited by Lynn Bryan and Kenneth Tobin, 313–326. New York: Peter Lang, 2018.
Herlihy, Mark. "Trying Creation: Scientific Disputes And Legal Strategies." *Science, Technology, and Human Values* 7, no. 40 (1982): 63–66.
Hodgson, Geoffrey Martin, and Thorbjørn Knudsen. *Darwin's Conjecture.* Chicago: University of Chicago Press, 2010.
Jenkins, Kristin P. "Evolution In Biology Education: Sparking Imaginations And Supporting Learning." *Evolution: Education And Outreach* 2, no. 3 (2009): 347–348. doi:10.1007/s12052-009-0158-8.
Keller, Evelyn Fox. *Secrets Of Life, Secrets Of Death.* Hoboken, NJ: Taylor and Francis, 1992.
Knain, Erik. "Ideologies In School Science Textbooks". *International Journal of Science Education* 23, no. 3 (2001): 319-329. doi:10.1080/095006901750066547.
Lakoff, George, and Mark Johnson. *Metaphors We Live By.* Chicago: University of Chicago Press, 1980.
Livingstone, David. "Science, Region And Religion: The Reception Of Darwinism In Princeton, Belfast, And Edinburgh." In *Disseminating Darwinism: The Role Of Place, Race, Religion, And Gender*, edited by Ronald Numbers and John Stenhouse, 7–38. Cambridge: Cambridge University Press, 1999.
Lopes, Sonia, and Sergio Rosso. *Bio*, vol. 1. São Paulo: Saraiva, 2015a.
Lopes, Sonia, and Sergio Rosso. *Bio*, vol. 2. São Paulo: Saraiva, 2015b.
Lopes, Sonia, and Sergio Rosso. *Bio*, vol. 3. São Paulo: Saraiva, 2015c.
Martins, Lilia. "A Historia Da Ciencia E O Ensino De Biologia." *Ciencia E Ensino* 5 (1998): 18–21.
Martusewicz, Rebecca A, Jeff Edmundson, and John Lupinacci. *Ecojustice Education.* New York: Routledge, 2011.
Mayr, Ernst. "The Philosophical Foundations Of Darwinism." *Proceedings of the American Philosophical Society* 145 (2001): 488–495.
Mayr, Ernst. *What Makes Biology Unique?* Cambridge, UK: Cambridge University Press, 2004.
Mayr. Ernst. "Introduction." In *On The Origin Of Species by Charles Darwin: A Facsimile Of The First Edition*, 1–6. Cambridge: Harvard University Press, 1964.

Mead, Rebecca, Momma Hejmadi, and Laurence Hurst. "Teaching Genetics Prior To Teaching Evolution Improves Evolution Understanding But Not Acceptance." *PLOS Biology* 15, no. 5 (2017): e2002255 doi: 10.1371/journal.pbio.2002255.

Miller, Kenneth, and Joseph Levine. *Biology*. San Francisco: Prentice Hall, 1991.

Mills, Susan, and Johm Beatty. "The Propsensity Interpretation Of Fitness." In Conceptual Issues In Evolutionary Biology, edited by Elliott Sober, 3–24. Cambridge: MIT Press, 2006.

Moody, David. "Evolution And The Textbook Structure Of Biology." *Science Education* 80, no. 4 (1996): 395–418.

Nadelson, Louis S. "Preservice Teacher Understanding And Vision Of How To Teach Biological Evolution." *Evolution: Education And Outreach* 2, no. 3 (2009): 490-504. doi:10.1007/s12052-008-0106-z.

Nehm, Ross H., Therese M. Poole, Mark E. Lyford, Sally G. Hoskins, Laura Carruth, Brent E. Ewers, and Patricia J. S. Colberg. "Does The Segregation Of Evolution In Biology Textbooks And Introductory Courses Reinforce Students' Faulty Mental Models Of Biology And Evolution?" *Evolution: Education And Outreach* 2, no. 3 (2009): 527–532. doi:10.1007/s12052-008-0100-5.

Numbers, Ronald. *The Creationists*. New York: Knopf, 1992.

O'Grady, Richard T. "Evolutionary Theory And Teleology". *Journal Of Theoretical Biology* 107, no. 4 (1984): 563–578. doi:10.1016/s0022-5193(84)80133-2.

O'Brien, Daniel Tumminelli, David Sloan Wilson, and Patricia H. Hawley. "'Evolution For Everyone'": A Course That Expands Evolutionary Theory Beyond The Biological Sciences." *Evolution: Education And Outreach* 2, no. 3 (2009): 445–457. doi:10.1007/s12052-009-0161-0.

Ontario Ministry of Education. *The Ontario Curriculum Grades 11 And 12: Science*. Toronto: Queen's Printer for Ontario, 2008.

Penteado, Pierre R., Karine F. Kavalco, and Rubens Pazza. "Influence Of Sociocultural Factors And Acceptance Of Creationism In The Comprehension Of Evolutionary Biology In Freshman Brazilian Students". *Evolution: Education And Outreach* 5, no.4 (2012): 589–594. doi:10.1007/s12052-012-0456-4.

Quessada, Marie-Pierre, Pierre Clement, Britta Oerke, and Adriana Valente. "Human Evolution In Science Textbooks From Twelve Different Countries." *Science Education International* 19, no. 2 (2008): 147–162.

Reis, Giuliano. "Making Science Relevant: Conceptual Change And The Politics Of Science Education." In *Reuniting Sociological And Psychological Perspectives*, edited by Wollf-Michael Roth, 233–241. New York: Springer, 2010.

Reis, Giuliano. "On And Off School Ground: A Discursive Approach To Science And Environmental Education." PhD diss., University of Victoria, 2007.

Reis, Giuliano, Nicholas Ng-A-Fook, and Lisa Glithero. "Provoking Ecojustice: Taking Citizen Science And Youth Activism Beyond The School Curriculum." In *Ecojustice, Citizen Science And Youth Activism*, edited by Michael Mueller and Deborah Tippings, 39–61. New York: Springer, 2015.

Reis, Giuliano, Christine Tippett, and Jeff Scott. "Pre-service Teachers' Perceptions Of Pedagogical Autonomy Through The Examination Of A Public Socio-scientific

Debate In The Media." In *Proceedings Of The European Science Education Research Association Biannual Conference* (ESERA), Nicosia (Cyprus), 2014.

Reiss, John. *Not By Design*. Berkeley: University of California Press, 2009.

Ritter, Bob, Richard Coombs, R. Bruce Drysdale, Grant Gardner, Dave Lunn, and Walter Samiroden. *Biology*. Scarborough: Nelson, 1993.

Robertws, Jonh. *Darwinism And The Divine In America: Protestant Intellectuals and Organic Evolution, 1859–1900*. Madison: University of Wisconsin Press, 1988.

Rudolph, John L., and Jim Stewart. "Evolution And The Nature Of Science: On The Historical Discord And Its Implications For Education". *Journal of Research In Science Teaching* 35, no. 10 (1998): 1069–1089. doi:10.1002/(sici)1098-2736(199812)35:10<1069::aid-tea2>3.3.co;2-1.

Ruse, Michael. *Charles Darwin*. Malden: Blackwell Publishing, 2008.

Scott, Eugenie. *Evolution Vs. Creationism*. Irvine: University of California Press, 2004.

Silva, Heslley, Graca Carvalho, Paloma da Silva, and Daiana Gibram. "The Perception Of The Hypothesis Of Intelligent Design In Brazil (Minas Gerais)." *Revista Conexao Ciência* 11, no. 1 (2016): 61–71.

Snyder, Vicky L., and Francis S. Broadway. "Queering High School Biology Textbooks". *Journal Of Research In Science Teaching* 41, no. 6 (2004): 617–636. doi:10.1002/tea.20014.

Sober, Elliott. *The Nature Of Selection: Evolutionary Theory In Philosophical Focus*. Chicago: University of Chicago Press, 1984.

StatCan (Statistics Canada). 2001. "Profile Of Disability Among Children." Accessed April 1, 2019. http://www.statcan.gc.ca/pub/89-577-x/4065023-eng.htm.

Sullivan, W. "Inheritance of Acquired Traits in Mice is Reported." *New York Times*, February 27, 1981 (A10). https://www.nytimes.com/1981/02/27/us/inheritance-of-acquired-traits-in-mice-is-reported.html.

Tomotoni, Joao, and Rodrigo Salvador. "Análise Do Conteúdo De Evolução Em Livros Didáticos Do Ensino Fundamental Brasileiro." *Pesquisa E Ensino De Ciências Exatas E Da Natureza* 1, no. 1 (2017): 5–18.

Trushell, John M. American Dreams of Mutants: "The X-Men—'Pulp' Fiction, Science Fiction, and Superheroes." *Journal of Popular Culture 38*, no. 1 (2004): 149–168.

United Nations. 2006. "Convention On The Rights Of Persons With Disabilities." Accessed April 1, 2019. http://www.un.org/disabilities/convention/conventionfull.shtml.

Willmersdorf, Pedro. "Casal Gay Recebe Carta Em Condomínio Da Zona Norte Do Rio: 'Gente De Cor E Afeminada.'" *Extra*, January 25, 2017. https://extra.globo.com/noticias/rio/casal-gay-recebe-carta-em-condominio-da-zona-norte-do-rio-gente-de-cor-afeminada-20819537.html.

Wilson, Edward. *The Social Conquest Of Earth*. New York: Liveright, 2012.

Woodfield, Andrew. *Teleology*. Cambridge: Cambridge University Press, 1976.

Conclusion

Kristin L. Cook and Alandeom W. Oliveira

As indicated in the first chapter, our main aim in this book was to examine how larger societal forces such as religion, media, and politics have shaped Brazil's educational landscape and impacted the teaching and learning of evolution within an increasingly polarized discourse in recent years. To this end, a number of educational scholars and practitioners, many of whom are based in Brazil, provided up-close and in-depth accounts of classroom-based evolution instruction, teacher preparation programs, current educational policies, and commonly used school curricula. Authors also presented information on Brazilian teachers' attitudes toward and understanding of teaching evolutionary theory, Brazilian students' attitudes toward and understanding of evolution, emergent (mis)conceptions of evolution as well as international comparisons of evolution acceptance and understanding in Brazil versus other countries. Together, these chapters paint a detailed and revealing picture of Brazil as a country that is currently experiencing an upsurge in creationist ideologies, a highly dynamic state of affairs that is creating complex challenges for science classrooms and teachers across the nation. Across the three sections of this book under which these ideas are housed, we see a nation navigating the complexity of multiple spheres of thought about evolution and its role in the K–12 and postsecondary curriculum. In this chapter, we synthesize the main themes that emerged out of the preceding chapters and highlight the implications (both theoretical and practical) for science teacher education. We then end by identifying areas in need of further research and practical consideration.

EMERGENT TRENDS

The chapters in part I that focused on evolutionary views, beliefs, and understandings of various social groups in different parts of Brazil highlighted the general populace's understanding and views about evolution and evolution education. Heslley Machado Silva's portrayal of the 390 adults surveyed in the study illustrates Brazilian respondents did not agree with the idea of exclusive teaching of creationism but appreciated the teaching of both creationism and evolution at school (though they recognized the value of Darwinian theory in scientific knowledge and advancements). They generally considered it possible to believe in God while simultaneously accepting Darwin's theory. Pedro Teixeira's study of evolution acceptance among high school students from two public high schools in Rio de Janeiro called attention to the importance of religious beliefs among students in the shaping of their ideas, and identified varied constraints that might hinder their understanding of the evolutionary theory. So, while the general population possess views that allow for both creationist ideas and the theory of evolution to coexist (i.e., to be harmoniously taught side by side), high school students who are learning about evolution against a backdrop of rising creationist ideologies actually have difficulty balancing their religious identities with scientific knowledge, which may negatively impact their understanding of evolution and its usefulness in science and society. These findings suggest that equal-time teaching, although theoretically possible and innocuous in the public's eyes (just a neutral and fair option), is in fact rather a highly problematic approach that can pose many challenges to learners who may experience it in the science classroom as being personally and epistemological dissonant (a source of confusion and added cognitive load). Its widespread endorsement by the public in Brazil and elsewhere as an effective educational approach to the teaching of evolution suggests a "folk theory" (Boudon, 1986; D'Andrade, 1987), a commonsensical and intuitive conceptualization of how students learn best that is in fact scientifically inaccurate (i.e., inconsistent with empirical research). Folk theories serve as cultural models (socially shared assumptions about a particular aspect of the world) that signal a kind of cultural wisdom. Handed down from authoritative figures and subtly promoted through the media, folk theories have been shown to constitute an influential source of bias and constraint for science education in general. As Windschitl (2004) writes,

> The science education community cannot discount the power of folk theories to influence instruction. . . . These folk theories, though mostly tacit, may have more influence on [science education] practice than authoritative documents (p. 508).

Likewise, the influential nature of folk theories about evolution education in Brazil is undeniable. A substantial portion of the public simply assume by default that creationism can be taught alongside evolution as a scientific theory without critically considering the implications and ramifications of such a proposal. Such an assumption cannot remain unquestioned or unchallenged. A good first step, we believe, is to raise the public's awareness of the epistemological inconsistencies, pedagogical challenges, and personal complications that will inevitably arise from this educational proposal through strategic public communication (debates, popular articles, etc.). Doing so might help push the public discourse beyond cultural wisdom toward critical consideration of evidence of what actually happens when creationism is allowed to enter the classroom under a scientific disguise. Science communicators have an important role to play in such an endeavor as potential partners for science educators. Their complementary expertise is likely to render efforts in this direction more successful.

Teixeira's call for the teaching of evolution to be oriented toward understanding its impact as a scientific theory without being imposed upon students as an absolute truth is commensurate with recommendations from other scholars. Because acceptance of evolution is connected to students' perception of evolution education, researchers recommend teachers ought to delineate the difference between science and other disciplines, address the distinction between acceptance of and understanding evolution, and discuss with students the laws pertaining to the teaching of evolution as well as the usefulness of the theory in modern-day applications (Donnelly et al. 2009; Cook 2009). As suggested by the National Center for Science Education and the National Research Council, comparing ideas that have two different rule systems (i.e., discussing evolution and religion as competing ideas) is not helpful in deepening student understanding of the nature of knowledge. Instead, it is essential to discuss how different disciplines (i.e., science, religion) have different ways of coming to know and justifying knowledge. Evolution is not only conceptually (and for some, religiously) challenging, but understanding and communicating about it requires an understanding of the epistemology of science.

The chapters in part II of this book that focused on evolution curriculum and instruction across a variety of Brazilian schools and university settings provided readers a glimpse of existing practices and pedagogies commonly found in Brazilian classrooms. Authors explored best practices, challenges, and affordances of curricular variations to teaching evolution, and the complexity of multiple perspectives surrounding evolution brought into classrooms by both students and teachers. Heslley Machado Silva and Eduardo Fleury Mortimer's portrait of high school biology teachers' avoidance of evolution and discomfort teaching the theory to students indicated a need for both pre-service and in-service teacher education that better supports new and experienced teachers in this area. New teachers are entering the school system without proper support or training on how to navigate the public controversy of the topic of evolution in the classroom while veteran teachers are facing new demands and discussions about evolution in curriculum for which they do not have clear guidance. Rita Tatiana Cardoso Erbs and Olma Karoline Cruz de Medeiros deepened our understanding of how teachers are grappling with teaching evolution in the high schools through their autobiographical research, which showed that many of the teachers, like those in Silva and Mortimer's study, took a relativistic stance toward evolution and religion. Especially in the area of macroevolution, teachers tried to marry ideas to avoid complexity or controversy. This left students with little to no understanding of the nature of science and what knowledge lay within the boundaries of science and what knowledge did not. Much of this work is commensurate with extant literature in evolution education on the need to improve biology teacher preparation programs in the country.

Several of the chapters in part II offered pedagogical strategies that targeted some of teachers' and students' pervasive and most resistant misconceptions in evolution. Beatriz Ceschim and Ana Maria de Andrade Caldeira provided specific pedagogical approaches to challenge teleological thinking, a common misconception about evolution despite scientifically known details about random mutation and how natural selection works at a genetic level. Gerda Maisa Jensen and Maria Elice de Brzezinski Prestes presented a teaching and learning sequence that centered on the history of biology to teach the theory of biological evolution and its main mechanism, the natural selection. The historical lens offered students an opportunity to look at competing claims and evidence used to justify them. Finally, Leonardo A. Luvison Araújo, Ronaldo A. Paesi, and Voltaire Paes-Neto illustrated ways to teach macroevolution, an important but highly controversial focal point in evolutionary theory that teachers and students find particularly problematic and at odds with their religious upbringings. Across of these chapters, we come to understand the importance of honoring student ideas and personally held worldviews while targeting misconceptions that deepen understanding of evolutionary history, the micro and macro processes of evolution, and the nature of science. These pedagogical strategies are in alignment with our own work, which has previously pointed to the effectiveness of dialogic approaches to evolution communication and instruction (Cook and Oliveira 2015). Rather than setting out to "rescue" students from their ignorance (i.e., taking a deficit-oriented, transmission-based stance on evolution aimed at informing them of the facts), it is more effective to adopt a lay expertise approach wherein evolution is respectfully presented without simply dismissing or excluding values, knowledge, and ways of knowing outside the scientific realm.

The above practices are also resonant with extant research on evolution instruction that supports inquiry-based and student-centered pedagogies. Most approaches in evolution education suggest encouraging biology teachers to elicit students' prior knowledge and construct experiences that allow students to examine and contrast those ideas with currently accepted scientific views (Khourey-Bowers 2006; Cook 2009). Researchers also call for the teaching of evolutionary theory to be explicitly linked to the teaching of nature of science (Anderson 2007; Cook et al. 2012) so that students can better understand the different epistemological considerations of different forms of knowledge and how each comes to be known.

The chapters in part III that presented international comparisons between Brazil and other countries helped readers discern the ways in which context and culture play a significant role in student learning of evolution, and highlighted the potential significance of issues raised and findings reported beyond Brazil itself. International comparisons of evolution education and public acceptance suggest a "global spread" of the creationist movement and raise concerns about potential growth of anti-evolution attitudes within public educational systems worldwide (Blancke and Kjærgaard 2016; Deniz and Borgerding 2018). Suggesting the emergence of a new generation of creationists in Brazil, scholars have pointed to a unique sociopolitical and educational context in a globalized era that allows for rapid worldwide travel of competing ideologies, unbounded by national borders (Oliveira and Cook 2018). Brazil is a country where religion is a strong component in the formation of people, and Brazilians in general possess a strong sense of faith in God as part of the cultural milieu. Additionally, there are changing demographics in Brazil (from predominantly Catholic to an increase in numbers of Protestant and Pentecostal/neo-Pentecostal denominations), which influence ideas and shape policy about the role of religion and science in schooling. In the classroom, teachers talk about "personal transformations" and "identity crisis" related to their teaching of evolution. The sociopolitical pressure affecting classrooms coupled with unclear messages from the Ministry of Education regarding the teaching of evolution makes Brazilians especially susceptible to undefined expectations and fuzzy parameters regarding what belongs in the science classroom and what does not.

Highlighted across the chapters in part III is the idea of how important schooling is for informing students of evolutionary theory regardless of religion. Nelio Bizzo's study of Brazilian and Italian student narratives from those living in fossil-rich environments show the different ways in which evidence regarding evolution can be conceived as well as how difficult the notion of deep time can be for students to conceptualize, even for those who live in areas where evidence literally surrounds them in the form of a plethora of fossil records in the nearby environment. Whether living in paleoisomorphic sites (where fossilized creatures are still living) or paleheteromorphic places (where fossilized organisms have become extinct), students did not implicitly acquire accurate knowledge about geological time. In fact, for many, the fossil record provided support for creationist ideas such as the occurrence of a biblical flood that extinguished many local organisms. Diving deeper into international comparisons of evolutionary understandings and beliefs, Graciela S. Oliveira, Helenadja Santos Mota, Ana Maria Santos Gouw, and Nelio Bizzo made comparisons among students in Brazil, Italy, and Ecuador (specifically, the Galapagos Islands) to elucidate the differences in knowledge and acceptance in the different countries. Specifically, it was found that Italian students were better informed about evolution than Brazilian or Ecuadorian students. Authors suggest this difference is due to differences in the educational systems. In Brazil, the authors state, young Brazilians may have had little or no contact with the idea of descent with modification in their school years. Authors point to the conceptual gaps among Brazilian students and suggest the issue rests most heavily in the curriculum presented to students in school (rather than to their ascribed religion). Finally, Giuliano Reis, Adam Brown, Delano Silva, and Ana Julia Pedreira illustrated how language present in classrooms, textbooks, and curricular documents in Canada and Brazil presented teleological misinformation that could distort views of evolutionary theory and could result in a faulty (and hierarchical) understanding of evolution as progress or improvement (e.g., social Darwinism). These international comparisons portray curricular supports in Brazilian classrooms as problematic and in need of clear guidance for learners with accurate scientific information.

FUTURE WORK

Public defense of the teaching of creationism by high profile politicians (Silva and Prado 2010) and controversial legal rulings such as the Brazilian Supreme Court's recent approval of "confessional religious teaching"—focused on a single religion—have blurred the idea of secularism in public schools. These pro-creationism upsurges have resulted in heated debates around the country about the separation of church and state, and the explicit inclusion of creationism alongside evolution in the curricular materials of nearly 1.5 million students in Brazilian public schools (Oliveira and Cook 2018). Without strong teacher guidance on what to teach, how to teach it, and the parameters within which one should teach evolution in the classroom, opportunities for conceptual and epistemological misunderstandings only increase.

While much attention has been given to students' conceptual development (e.g., students' and teachers' ideas and conceptions) and acceptance of evolution (Donnelly et al. 2009), students' and teachers' evolution communication itself (communicative forms, strategies, and skills) remains to be examined in more depth. This type of exploration provides glimpses into the nuances of teaching and strengthens pedagogical recommendations for educators. Teachers treading lightly around the topic of evolution or assuming that students implicitly glean the nuances of the nature of science is simply inadequate. Much of our work has focused on the importance of communicating evolution and guiding teachers with clear content expectations but also the ability to communicate and support students' communication about evolution (Oliveira et al. 2011; Cook and Oliveira 2015; Oliveira and Cook 2016). We maintain that a communication-centered approach to evolution instruction encourages students to extend beyond the walls of the classroom and foster more meaningful classroom experiences that can promote higher levels of student engagement, conceptual understanding, and development of essential science communication skills. In addition to developing a scientific understanding of evolution, students need to become capable of expressing their understandings to others. Student attainment of *controversial communicative competence* (Cook and Oliveira 2015)—the ability to competently communicate about socially controversial science topics—in order to fully flourish as informed citizens capable of productively negotiating scientific, public, and classroom spheres. As an essential practice of science, communication is central to students' fulfillment of their democratic roles.

However, pedagogically equipping teachers with necessary classroom supports is certainly not enough. In the political, religious, and social spheres that surround a classroom, teachers and students find themselves confronted with mixed messages and mired in propaganda. In Brazil and elsewhere, polarization has arisen from the reaction of religious conservatives to rapidly changing public policy and diverse value systems. Particularly in Brazil, culture wars are most prominently between Catholics vs. Evangelicals and between religious vs. secular groups (Smith 2019). In what Smith describes as a "clergy-driven culture," religious leaders have power to select candidates and substantial influence over their platforms. Perceived threats to religious traditions and values has led to the "mobilizing of the People of God," that is, intense engagement of religious groups in political activism. This has contributed to intense polarization on many issues, including the debate over what should be taught in science classrooms and what is meant by secular education. As we have learned from other nations' histories, the importance of clear messages from political and school leaders as well as the selection of curriculum and texts (through postsecondary) that undergird those messages cannot be overstated. Moreover, it will be critical for the science education community in Brazil to find ways to counter the "culture of agonism" (Tannen 2002) prevalent in Brazilian political discourse. In an agonistic culture, deliberations are approached as a polarized "battle" between two warring camps, being characterized by excessive combativeness/adversarialness and disruptive disagreement (vitriolic attacks, sarcastic innuendo, and even name-calling). Discussing the educational value of evolution in such a cultural context is unlikely to be productive as agonistic opposition for the most part obfuscates democratic negotiation and can completely hamper collaboration and communication. Such an endeavor demands a healthier and more positive sociopolitical atmosphere that favors cooperation over combat, respect for others over personal offense, and intellectual reason over political games. It is our hope that the present book will assist biology educators in Brazil and in other countries with this daunting task and provide guidance on how to help different societal groups more clearly recognize the paramount importance (democratic, societal, and scientific) of teaching and learning evolution in public schools, regardless of religion, or political affiliation.

REFERENCES

Anderson, Ronald. 2007. "Teaching the Theory of Evolution in Social, Intellectual, and Pedagogical Context." *Science Education* 10: 664–677.

Blancke, Stefaan and Peter Kjærgaard. 2016. "Creationism Invades Europe." *Scientific American*. Retrieved from https://www.scientificamerican.com/article/eurocreationism/

Boudon, Raymond. 1986. *The analysis of ideology* (M. Slater, Trans.). Chicago: University of Chicago Press.

Cook, Kristin. 2009. "A Project-Based Evolution Unit for High School: Teaching Content through Application." *The American Biology Teacher* 71: 95–100.

Cook, Kristin and Alandeom Oliveira. 2015. "Communicating Evolution: An Exploration of Students' Skills in an Essential Practice of Science." *Electronic Journal of Science Education* 19, no. 5: 1–23.

Cook, Kristin, Gayle Buck, and Meredith Park Rogers. 2012. "Preparing Educators to Teach Evolution in a Project-based Approach." *The Science Educator* 21: 44–56.

D'Andrade, Roy. (1987). A folk model of the mind. In D. Holland & N. Quinn (Eds.), *Folk Theories in Language and Thought* (pp. 114–148). Cambridge, England: Cambridge University Press.

Deniz, Hasan and Lisa Borgerding. 2018. "Evolution Education around the Globe." New York: Springer.

Donnelly, Lisa, Massa Kazempour, and Amir Amirshokoohi. 2009. "High School Students' Perceptions of Evolution Instruction: Acceptance and Evolution Learning Experiences." *Research in Science Education* 39, no. 5: 643–660.

Khourey-Bowers, Claudia. 2006. "Structured Academic Controversy: A Peaceful Approach to Controversial Issues." *The American Biology Teacher* 68, no. 5: 43–47.

Oliveira, Alandeom, Kristin Cook, and Gayle Buck. 2011. "Framing Evolution Discussion Intellectually." *Journal for Research in Science Teaching* 48, no. 3: 257–280.

Oliveira, Alandeom, and Kristin Cook. 2016. "Student Visual Communication of Evolution." *Research in Science Education* 46: 1–20.

Olivera, Alandeom, and Kristin Cook. 2018. "The Rise of the Creationist Movement in Brazil." In L. Borderding and H. Deniz (Ed.) *Evolution Education around the Globe*. New York: Springer.

Silva, Heslley, and Isabelle Gonçalves de Oliveira Prado. 2010. "Creationism and Intelligent Design: Presence in the Brazilian Educational Policy." *Procedia Social and Behavioral Sciences* 2: 5260–5264.

Smith, Amy. 2019. "Religion and Brazilian Democracy: Mobilizing the People of God." Retrieved from: http://amyericasmith.org/religion-and-brazilian-democracy/

Tannen, Deborah. 2002. "Agonism in Academic Discourse." *Journal of Pragmatics*, 34, 1651–1669.

Windschitl, Mark 2004. "Folk Theories of 'Inquiry': How Preservice Teachers Reproduce the Discourse and Practices of an Atheoretical Scientific Method." *Journal of Research in Science Teaching*, 41(5): 481–512.

Index

About the Editors and Contributors

Leonardo A. Luvison Araújo is a PhD candidate in education at the Universidade Federal do Rio Grande do Sul, Brazil. He earned a licensure in biology (2012) and a master's degree in genetics and molecular biology (2015) at the Universidade Federal do Rio Grande do Sul. He has taught continuing education courses about evolution to teachers and undergraduate biology students in Brazil. His research interests include evolutionary biology, evolution teaching, and history and philosophy of biology.

Nelio Bizzo is a biologist and educator, full senior professor of science education at the Faculty of Education, São Paulo University, Brazil. His work is related to biological education, with a special focus on the learning and teaching of biological evolution in Brazil and other countries. He is the scientific coordinator of the research nucleus on Education, Epistemology of Evolution and Outreach (EDEVO-Darwin/USP), and of a five-year-long project, financed by FAPESP, focusing educational aspects of Brazilian biodiversity (Biota-Educação). His research group has developed methods on sociocultural grounds, combining qualitative and, more recently, quantitative methods. The History of Biology is one of his major research subjects, with a special focus on the development of Charles Darwin's works, having developed some research with primary sources at Cambridge University Library (Manuscripts Room). Nelio is a Researcher of the National Council of Research (CNPq, 1A) in Brazil since 1995, was elected fellow of the Royal Society of Biology (London), and has spent some time as visiting professor in Italy (Verona and Padova Universities). He worked as an expert member of the Intergovernmental Panel on Biodiversity and Ecosystems Services (2014–2019).

Adam Oliver Brown is professor in the Department of Biology at the University of Ottawa. Trained as a pollination ecologist (PhD, Laval), he also has degrees in biology and environmental science (BSc, UWO) and ecology and evolution (Hons. Standing BSc, UWO). Since joining as a faculty member at uOttawa, Dr. Brown has developed an interest in science education and science communication and has spent much of the past decade-plus focusing his research on the scholarship of teaching, learning and communicating science. He is the university's first cross-appointed professor between the Faculties of Science and Education, an affiliate member of the uOttawa Institute for the Environment and an associate member of the world-renowned science communication team at Montreal's McGill University, the Office for Science and Society. Adam has also been a performing artist for most of his life, having entertained international audiences with music, dance, and theater for over three decades.

Ana Maria de Andrade Caldeira is associate professor of science education at the São Paulo State University (Unesp), School of Sciences, Bauru. She earned a master's degree in genetics at the University of São Paulo (1984) and a PhD in education at São Paulo State University (Unesp) (1998). Accredited by the postgraduate program in Education for Science, she directs masters and doctoratal students in the didactic and epistemology of biology.

Beatriz Ceschim is a PhD student of science education in the postgraduate program in Education for Science at the São Paulo State University (Unesp), School of Sciences, Bauru. She earned a master's degree in science education (2017). Her research interests include didactic and epistemology of biology.

Kristin L. Cook is associate professor of science education at Bellarmine University and serves as the associate dean of the School of Education. She received her doctoral degree at Indiana University in curriculum and instruction, specializing in science education and environmental sciences. A former high school biology teacher, Dr. Cook has served as a professional developer and educational consultant for elementary, middle, and secondary school innovations. Dr. Cook's research focuses on engaging students and teachers with the community of science through the exploration of socio-scientific inquiry and transdisciplinary instruction.

Rita Tatiana Cardoso Erbs is associate professor at the Federal University of Goiás–Regional Catalão. Rita's interests include teaching, conducting research, and working with community outreach. She holds a bachelor's degree in psychology from PUC-RS (1993), a masters' degree in education from

PUC-RS (2001), and a PhD in education from PUC-RS (2010). Her doctoral study centered around Buddhist wisdom and teacher training for a culture of peace. She has experience in education, clinical psychology, and work psychology with an emphasis on autobiographical research and life stories. She mainly works in the following areas: distance education, preparation of didactic materials, teacher training, evaluation, mental health, and work. At the undergraduate level, she teaches psychology of education to students with different majors: biological sciences, physical education, psychology, pedagogy, among others. At the graduate level, she integrates the research traditions of educational policies, history of education. and research (auto) biographical.

Ana Maria Santos Gouw is professor at the Federal University of Sao Paulo, Brazil. She holds a master's degree and a PhD in education from the University of São Paulo, Brazil (2013). Her main research interests are science education research and teaching of biological knowledge.

Gerda Maisa Jensen is a retired secondary school teacher in the city of São Paulo, Brazil. She holds a master's degree in history of science from the Pontifical Catholic University of São Paulo (PUC/SP) (2008) and a PhD in biological sciences/genetics from the Institute of Biosciences of the University of São Paulo (IB/USP) (2016). She taught science classes to students from private and public schools where she also coordinated professional development programs for teachers. Her research interests lie in the history of biology and in the introduction of the history of biology in science education through inquiry-based teaching for students of the Young and Adults Education Program.

Olma Karoline Cruz de Medeiros majored in biological sciences (2003) and earned a master's degree in science and mathematics education from the Federal University of Uberlândia (2017). She has fifteen years of experience working with elementary, high school, and college students both in the private and public systems. She conducts studies in the area of teacher training and science teacher. She is the author of two books entitled *Biopráticas: Experimental Activities* and *Biomúsicas: Parodies on Biological Contents*.

Eduardo F. Mortimer graduated in chemistry (1980) and has a master's degree in education (1988) from the Universidade Federal de Minas Gerais, Belo Horizonte, Brazil. He got his PhD (1994) from the Universidade de São Paulo, Brazil. He is professor of science education at the Universidade Federal de Minas Gerais and researcher at CNPq, Brazil. His research interests include classroom discourse and conceptual profiles. His publications include *Meaning Making in Secondary Science Classroom* (2003) with Phil Scott, and *Conceptual Profiles: A Theory of Teaching and Learning Scientific Concepts* (2014) with Charbel N. El-Hani.

Helenadja Santos Mota is professor at the Federal Institute Catarinense, Brazil. She holds a master's degree in history, philosophy, and science teaching from Federal University of Bahia, Brazil (2009) and a PhD in education from the University of São Paulo, Brazil (2013). Her main research interests are science education research and teaching of biological evolution.

Voltaire Dutra Paes Neto is a PhD student in geosciences at Universidade Federal do Rio Grande do Sul (UFRGS) at Porto Alegre. He earned a master's degree in geosciences at the same institution. He taught biology classes (2011–2017) in a popular course entitled "PEAC-Citizen Alternative" and worked on the outreach promotion of paleontology and biological evolution to young students, teaching courses, and collaborating on book chapters. His research interests include vertebrate evolution, cladistics, and the teaching of paleontology.

Alandeom W. Oliveira is associate professor of science education at the State University of New York at Albany. He earned a master's degree in science education at Southeast Missouri State University (2002) and a PhD in science education at Indiana University Bloomington (2008). He has taught science education courses to teachers in Brazil and the United States and has coordinated multiple professional development programs for school teachers, including Science Modeling for Inquiring Teachers Network and Technology-Enhanced Multimodal Instruction in Science and Math for English Language Learners. His research interests include cooperative science learning, inquiry-based teaching, and classroom discourse.

Graciela da Silva Oliveira is professor at the Federal University of Mato Grosso, Brazil. She holds a master's degree (2009) and a PhD (2015) in education from the University of São Paulo, Brazil. Her main research interests are science education and culture research.

Ronaldo Antonio Paesi is a PhD candidate in science education at the Universidade Federal do Rio Grande do Sul, Brazil. He earned a licensure in biology (2014), a baccalaureate in biology (2016), and a master's degree in animal biology (2016) at Universidade Federal do Rio Grande do Sul. He is a professor at the Instituto Federal de Educação Ciência e Tecnologia Rio

Grande do Sul. He has taught continuing education courses about evolution to teachers and undergraduate biology students in Brazil. His research interests include evolutionary biology and evolution teaching.

Ana Julia Pedreira is professor of science education at the University of Brasília (UnB), Brazil. She earned a master's degree in ecology at the University of Brasilia (1998) and a PhD in education at the same university (2016). Since 2018, she coordinates the Biology Pedagogical Residence, a project where student teachers spend eighteen months teaching in local schools. Her research interests include teaching strategies, science textbooks, and teacher education.

Maria Elice de Brzezinski Prestes is associate professor of history of biology and biology teaching at the Institute of Biosciences of the University of São Paulo, with a previous eleven years of experience teaching biological science in high schools. She holds a master's degree in environmental sciences (1997) and a PhD in education (2003), both at the University of São Paulo. Her research interest in the history of science includes the seventeenth- and eighteenth-century observational and experimental studies on living beings. She also supervises research on the use of episodes of the history of biology in science and biology education.

Giuliano Reis is associate professor of science education at the Faculty of Education and also an affiliate member of Institute for the Environment at the University of Ottawa (Canada). He holds an MA in education (University of Brasilia, Brazil) and a PhD in science education (University of Victoria, Canada). His main research interest is in the study of language use in/for science and environmental education in various contexts. He is married to Juliana Reis and together they have two daughters: Ana-Julia and Maria-Luiza Reis.

Delano M. S. Silva is associate professor at the University of Brasilia, Brazil. He earned a master's degree in ecology at University of Brasilia (1998) and a PhD degree in ecology at the same university (2005). He is a practicum supervisor and delivers professional development courses to teachers in Brasilia on subjects like practicum, didactic resources, and inquiry-based learning. His research interests include science teaching, inquiry-based learning, teacher identity, and teacher pedagogical content knowledge.

Heslley Machado Silva is full professor and project coordinator of the pedagogical residency program of the University Center of Formiga/MG (UNIFOR/MG), having coordinated courses in biological sciences. He has been the coordinator of the project, Biology Teachers' Conceptions of Three Latin American Countries about Humans' Place in Nature and the Human Mind, by the University Center of Formiga/MG and University of Oxford, and supported by the John Templeton Foundation (2016). His research focuses on issues related to the teaching of evolutionary biology in Latin America. He did his postdoctoral internship in education at the University of Minho, in Braga, Portugal (2016); his PhD in education at the Federal University of Minas Gerais (2015), Brazil; his MSc in education at the Federal University of Minas Gerais (2000); and his graduation in biology and science at the University Center of Formiga/MG (1995). He also has a specialization in biotechnology at the Federal University of Lavras/MG and another specialization in molecular and cellular biology from the Federal University of Ouro Preto/MG. He has experience in general biology, with emphasis on embryology, genetics, evolution, and zoology. He is coeditor of the journal *Conexão Ciência* (*Science Connection*).

Pedro Teixeira is assistant professor at the Department of Education of Pontificial Catholic University of Rio de Janeiro (PUC-Rio, Brazil) where he leads Diversias Research Group on Diversity, Education, and Controversies. His main research interests are in diversity, religion, science, biology education, and interculturality. He earned a PhD degree in human sciences: education at PUC-Rio, with a one-year study period at the Institute of Education of the University College London (UCL, UK) under the supervision of Dr. Ralph Levinson. He is associate researcher of REDISCO: Religions, Discriminations en Milieu Scolaire [Religions, Discriminations at School Environment] (Université Lyon 2, France). He also has a master's in education at PUC-Rio and has been a science and biology teacher for six years.

Lightning Source UK Ltd.
Milton Keynes UK
UKHW020951221019
352058UK00005B/382/P

9 781793 601483